Don't Throw This Away!
The Civil Engineering Life

Other Titles of Interest

Changing Our World: True Stories of Women Engineers
by Sybil E. Hatch (ASCE Press). Real-life stories about the lives and careers of hundreds of women engineers, celebrating their contributions to every aspect of modern life.

Designed for Dry Feet: Flood Protection and Land Reclamation in the Netherlands
by Robert J. Hoeksema (ASCE Press). An engineer's account of Holland's unique challenges in water control and management across several centuries.

Engineering Legends: Great American Civil Engineers
by Richard G. Weingardt (ASCE Press). Sketches of the lives and achievements of 32 great American civil engineers, from the 1700s to the present.

Engineering Your Future, Second Edition
by Stuart G. Walesh (ASCE Press). Supplement to the technical preparation of engineers and other professionals, providing valuable advice and instruction crucial to success in today's world.

Managing and Leading: 52 Lessons Learned for Engineers
by Stuart G. Walesh (ASCE Press). Useful ideas and fundamental principles for engineers to improve their management and leadership skills.

Tipon: Water Engineering Masterpiece of the Inca Empire
by Kenneth R. Wright (ASCE Press). An appreciation, for modern engineers and casual tourists alike, of the Inca civilization through the great engineering works they left behind.

Don't Throw This Away!
The Civil Engineering Life

Brian Brenner, P.E.

Library of Congress Cataloging-in-Publication Data
Brenner, Brian R.
Don't throw this away! : the civil engineering life / Brian Brenner.
p. cm.
ISBN-13: 978-0-7844-0888-9
ISBN-10: 0-7844-0888-2
1. Civil engineering. I. Title.

TA155.B74 2007
624—dc22

2006026264

Published by American Society of Civil Engineers
1801 Alexander Bell Drive
Reston, Virginia 20191
www.pubs.asce.org

Cover illustration by Garry Nichols—Images.com.

ISBN 13: 978-0-7844-0888-9
ISBN 10: 0-7844-0888-2
Manufactured in the United States of America.

For Lauren,
who took the Bridge Test
and passed with flying colors

Contents

Foreword, *Paul C. Taylor, P.E.* **ix**
Acknowledgments **xi**

Brian's Bridges **1**
Life Insurance **4**
The Twister **7**
Engineering Fashions **11**
Who Likes the Chocolate? **14**
The Maze **17**
The Ronald Reagan Room **20**
Bringing Out the Inner Civil Engineer **23**
The Baby Sitter-In-Law **26**
Reston Town Center **29**
Don't Throw This Away, I **32**
Don't Throw This Away, II **34**
Don't Throw This Away, III **37**
Don't Throw This Away, IV **40**
An Ideal Geotechnical World **43**
Infrastructure and Coming of Age **45**
Learning the Expanding Body of Knowledge **48**
The Road Not Built **51**
Build It and They Will Come **55**
New Car **58**

Acronyms and the Explosion of Useless Data (AEUD) **61**
The Quacking Moment **65**
Fish **69**
The Bridge Tour **72**
Fred Retires **76**
The Sky Bridges and Malls of Minneapolis **78**
Raising the Bar **81**
Vegetarian Nerds Watching the Super Bowl **84**
My New Cell Phone **88**
Hamsters Gone Wild **91**
A Comparison of Dilbert and Wally **95**
The Discovery of Pluto **98**
What's New on the Xway **101**
Opryland **105**
The Last Game at Foxboro **108**
The Zucchini Story **112**
Moss on the Median **116**
First Class **120**
After All, It's a Small World **123**
What Happened to Nantucket? **126**
The Way Things Are **129**
The Forest and the Trees **132**
Mass MoCA and the Hoosac Tunnel **134**
The Trail Ridge Road **137**
Encino Engineer **140**
Aberaeron **142**

Publishing Credits **146**
About the Author **148**

Foreword

Everyone knows an engineer can't be a good writer. Everyone knows an engineer can't ever be funny. Everyone knows an engineer can't be introspective. Everyone knows an engineer has no insight into the world outside her or his cubicle.

Then everyone doesn't know Brian Brenner. And certainly everyone doesn't know the Brian Brenner I know.

I first heard (or, more significantly, read) the name Brian Brenner when it kept appearing in the byline of strange articles in my company's "technical" newsletter. Those articles were technical, yes, but also interesting to read, thoughtful, often funny, and usually related the "technical stuff" to the "real world."

Was Brian Brenner the *nom de plume* of some English major who'd infiltrated the technical network? Well, I would soon find out, courtesy of the American Society of Civil Engineers, no less. Before long, I found myself sitting in a committee meeting face to face with a guy calling himself Brian Brenner. At first, he seemed every bit another of us clueless engineers. But when he volunteered (yes, volunteered) to write up lengthy minutes of the committee's doings, I got suspicious.

The next few years of committee service proved to me conclusively that (a) Brian Brenner is really an engineer, (b) Brian Brenner is a good writer, (c) Brian Brenner has some interesting thoughts about life, (d) Brian Brenner pays attention to what's going on in the world around us, and (e) Brian Brenner can be funny sometimes. It's the funny times I remember most, as they came when Brian and I engaged in one pastime we both enjoy very much—drinking good beer. Come to think of it, there was plenty of introspection and insights into the wider world as we imbibed in places like Fort Collins, Tampa, Seattle, Baltimore, and, of course, Reston.

Now everyone can get to know Brian Brenner through this unique collection of essays that just may be a first for civil engineering.

Why do I say it's a unique collection? Perhaps because the subjects range so far and wide. From urban form and suburban sprawl to tips on an organized, stack-free life (boy, can I use *those* tips). From electric power for Cape Cod to twisters in the Midwest. And from Hershey to Nantucket to a village in Wales.

Why do I say this may be a first for civil engineering? Perhaps because it speaks to our inner engineer, prompting us not to expect each design to be perfect and evoking the feelings we get on opening day when we see our paper-napkin idea at full scale. Or maybe because his message is that "envisioning … the future … is good for civil engineers to do … because no one else seems to be doing it." And surely because the collection consistently celebrates "civil engineering glee," reminding us that every day we are revisiting our days of blocks and Brio trains and sandcastles.

I have just one objection to Brian's *oeuvre*. At one point, he alleges that his "sarcasm is pretty good." That may be true but I wish he hadn't so blatantly betrayed his East Coast myopia in choosing as a principal target the place I have called home for decades. In a poignant reflection on how the world "is increasingly homogenized and the same," Brian predicts that "soon every place will look like Southern California." (If the weather went along with it, maybe that wouldn't be so bad.) Come to think of it, there's so much sarcasm in this collection that you too may find your home town or your specialty practice or your favorite beer lampooned as well.

Dig in and find out. I think it's worth it.

Paul C. Taylor, P.E.
Culver City, California

Acknowledgments

This book is full of hoary clichés, and here's another: having a book published like this is a dream come true for me. When I've described what I was working on to colleagues, I said that the book was the written version of me. It takes a village to create a book, and I want to thank my friends and colleagues for being the villagers:

Terrific editors Lorraine Anderson, Willa Garnick, and Gian Lombardo have provided great comments and improvements for my writing over the years. What luck I've had to work with the excellent ASCE Press editor Betsy Kulamer, who worked on this manuscript and greatly improved it.

I gave some of my Tufts students an assignment to write an essay, "The Students Get to Be in the Book." Unfortunately, the students didn't get to be in the book because the essay didn't make the final cut, but here's thanks to contributors Christina Loulakis, Norm Quach, Mike Diminico, Heather Shields, Danny McGee, Stephanie Fowler, Allison McCarthy, Dave Czulada, Aaron Levine, Brian Mackey, and Jess Pransky. Also thanks to Cornell student Tucker Moffat for being the Tufts civil engineering class mascot. Best wishes to the Tufts CE graduating Class of 2006—you are great and the world is going to be a lot better when you're done. Thanks to all my colleagues at Tufts: Professors Masoud Sanayei, Lewis Edgers, Chris Swan, Lee Minardi, Luis Dorman, Eric Hines, Laurie Baise, Dean of Engineering Linda Abriola, and other members of the civil and environmental engineering department.

I have had the good fortune to work with many exceptional engineers at Parsons Brinckerhoff, too many to list here. Special thanks to the *PB Network* crew, John Chow, Gordon Clark, and Laurie Ludwin. *PB*

Network is the corporate technical magazine of Parsons Brinckerhoff. The magazine is available at http://www.pbworld.com/news_events/publications/network. Original versions of some of the essays in this book can be found there.

I have been fortunate to participate in committees for both ASCE and the Boston Society of Civil Engineers Section (BSCES), again with a very long list of people to whom I'm indebted for help and support over the years. Thanks to colleagues at the *Journal of Professional Issues in Engineering Education and Practice,* editor Norb Delatte, Paul Taylor, Jerry Rogers, Bill Lawson, Norm Dennis, Dennis Truax, Amarjit Singh, the great corresponding editors, and ASCE Journals Director Johanna Reinhart and Managing Editor Jackie Perry. In Boston, we are proud of the BSCES, the oldest U.S. engineering society, which even predates ASCE by a few years. Many thanks to many BSCES colleagues, including Ali Touran, Cindy Chabot, Anni Autio, David Manugian, Yanni Tsipis, Abbie Goodman, and Reed Brockman.

I'll need a few more pages to thank the friends, but two made it into the book: my technological friend, Seth Frielich, and the engineer's engineer's engineer, Aren Horowitz. Steve Binney's attention is directed to the essay beginning on page 69. Thanks to British civil engineer Paul Jackson for his comments and additions to the essay, "Aberaeron." My wry and endlessly talented son, Daniel, has provided hours of bemusement and the opportunity to use big words like "bemusement." He is going to the University of Maryland and has stuck up my car with UM bumper stickers, requiring interesting explanations at Tufts. Also a great source of material is my brilliant, literate daughter, Rachel, who among many other things introduced us to hamsters. Putting it all together, a great, additional, extra-special thanks to my beautiful wife, Lauren, who gets to experience me 24/7. If you think the book version is a challenge, you can check with her about dealing with the live commodity.

Brian's Bridges

Who creates a bridge, I wonder. A bridge starts out as just air. There is a space over the river or highway where the bridge will cross. Many years later, the plans are drawn, and construction equipment appears at the site. For those not familiar with the process, the design and construction of a great bridge must all seem a little mysterious. Somehow the piers appear in the harbor, and towers rise out of the water. If it's a suspension bridge, cables are strung from the top, and pieces of deck are lifted to make the span. The whole thing is connected, paved, and blessed, and on the appointed day, the signs are unsheathed, the lights turned on, the speeches made, and the wondrous structure assumes its place in the pantheon of anonymous infrastructure.

When I was a toddler, my parents drove to Grandma's house, and we passed the site of the Verrazano-Narrows Bridge. This magnificent structure was under construction across the Narrows between Brooklyn and Staten Island. The partially built towers leapt out of the water on their artificial islands. Huge cranes hoisted the steel, and giant spools strung the cables. This made a big impression on my four-year-old eyes, and soon I was building suspension bridges with my toy blocks. I received an early structural education in this way, figuring out the best way to anchor the string cables. After experimenting, I used heavy pieces at both anchorages to resist the tension. I mimicked the construction process, using my toy boats to transport the deck sections in the "harbor" and raise them into place. When the structure was done, I contacted the owner and showed him my work.

"What do you think of my bridge, Dad?" I said to the client.

"Nice job," he responded.

If it was a special creation—and the suspension bridges usually were—the other client allowed it to remain overnight before cleanup the next day. There was a streetlight outside my bedroom window, and I remember going to sleep with the dim glow reflecting off the string cables.

It appeared that I was a budding engineer, so my mother wrote to the governor and got tickets to the opening ceremony of the Verrazano Bridge. A big crowd gathered at the Staten Island toll plaza on a bright and sunny November day for long speeches and pontification. Most of the participants didn't have much to do with the bridge's design and construction, other than to show up for the dedication. Robert Moses, the chairman of the Triborough Bridge and Tunnel Authority, pointed to the chief engineer, Othmar Amman, but didn't mention him by name.* This was as close as the speakers got to acknowledging the engineers, since Amman himself was not invited to the podium. I still have the Opening Day brochure and commemorative stamps in my bridge scrapbook. My mother wrote on the front of the scrapbook, "Brian's Bridges."

A great bridge is the product of the imagination and sweat of hundreds of people. Maybe this is why a bridge is not easily identified with individuals and why a bridge's creation and birth seem anonymous. Very few of the structures are named in honor of the people who created them. In fact, most bridge names honor people who had nothing to do with the work. In the case of the Verrazano Bridge, naming the structure after its creators would have required a very long name. Society expects that engineers and constructors will fade into the background, like the bridges. The builders can be proud of their creations but must be satisfied that the symbolic act of naming, the official recognition of the creation, will be transferred to someone else.

In March 2003, my father visited Boston. It was a warm afternoon after a long, bitter winter. We went to visit the Zakim Bridge, just days before the first part of its staged opening. The Zakim Bridge stood tall and sleek in the middle of dowdy, old downtown Boston. Hundreds of thousands had watched the structure appear from nothing, with its futuristic concrete pylons and slender cables strung to the deck one piece at a time. Upon completion, the structure quickly became an infrastructure icon for the city, with its image appearing on bank advertisements,

*Caro, Robert. (1975). *The Power Broker: Robert Moses and the Fall of New York,* Vintage, New York.

at the beginning of newscasts, and in dozens of unrelated publications. On the Fourth of July, the blue tower lights were supplemented by a red glow at night, so with the white cables, the bridge was patriotic, a giant cable-stayed American flag.

We stood by the bridge. I had little direct involvement with this bridge design, but I said to my father:

"What do you think of my bridge, Dad?"

"Nice job," he replied.

Life Insurance

I applied for a supplemental life insurance policy. In order to approve my request, the insurance underwriter interviewed me. Among other requirements, the company wanted to verify that I was still alive. This was considered a baseline condition for the policy. The interview was conducted by telephone. I don't have an exact transcript, but it went something like this:

Underwriter: Hello. I am calling to interview you for your life insurance policy application.

Me: OK.

Underwriter: It will take from five to seven minutes.

Me: OK.

Underwriter: Please state your name, address, yada yada yada.

I gave my name and address and responded to the yada yada yada part.

Underwriter: Have you had one of the following conditions: heart disease, lung disease, cancer, AIDS, neurological symptoms, difficulty in breathing?

Me: No.

Underwriter: Have you ever lost your liver?

Me: No.

Underwriter: Are you insane?

Me: No.

The interview proceeded along these lines for a while. The underwriter was trying to establish risk categories. For example, if I was insane, then the insurance company probably had a study showing that I might jump off a building or take some other insane action. This would increase the risk for payout of the policy. Their business model was based on having clients who would only pay in with no risk of paying out. Ideally, all clients would be people in perfect health who were also immortal.

Then the questioning started to get peculiar.

Underwriter: Have you ever participated in bungee-cord jumping? Paragliding? Ice-climbing up shear cliffs without wearing gloves? Any extreme and dangerous sports that we should know about?

I thought about this one before responding. The truth was that I never participated in any extreme sports at all. Even when I went skiing, it was only down the bunny trail with my young daughter and some first-grade kids. Although this activity was perfectly safe for me, the first-graders were in danger when I used their bodies to stop my forward motion. I suppose that should add risk to their life insurance policies. But not mine.

Yet, for the sake of argument, suppose that I actually was a semiprofessional bungee-cord jumper? How would the underwriter ever find out the truth? Maybe there would be a smoking gun video of me at the semiprofessional bungee-cord jumping competition. I had seen on TV such a contest conducted off the New River Gorge Bridge in West Virginia. (If I was going to do it, it would be off a bridge—and not just any bridge, but the world's second-longest arch bridge span.)

I said to the underwriter: "Could you please repeat the question?"

For a moment, the conversation stopped. Probably no one had ever asked for that question to be repeated. What was going on here? The underwriter was undoubtedly thinking, what does this guy have to hide? Hang-gliders are charged a 70% extra insurance premium because of the danger. If the underwriter operated on commission, then suddenly the meter was on and running. Ka-ching!

The underwriter repeated the question.

I responded, "No."

For the rest of the call, the underwriter was a little bit annoyed. There were no smoking guns here. No severe health problems, no unseemly risks. The interviewee, me, was low-risk and boring. Then we got to the crux of the matter.

Underwriter: What is your profession?

Me: Civil engineer.

I could almost hear the balloon deflate on the other end of the phone (Thrpppppppptttt!!!). The underwriter was thinking, "Ah, a civil engineer." On the actuarial tables, the only people less risky to insure than civil engineers are life insurance underwriters, and then only slightly. The same qualities that make civil engineers less glamorous than, say, movie stars or brain surgeons, make them good insurance risks. For all our dullness and lampooning in *Dilbert,* for all our relatively low status in society, we should at least expect in return to have low life insurance rates.

I thought of the times that my wife and I visited time-share resorts. These former motels offered fabulous prizes when you visited them. All you had to do was sit for a mere 90 minutes or so while they tried to sell you a week at their facility. It sounded innocent enough, and we liked the part about winning a fabulous prize. But in reality, the salesmen had cooked up a clever promotional scheme not intended for the faint of heart. The innocent 90-minute tours featured extreme, high-pressure sales pitches, where the victims sat indebted to the perpetrators because of the potential for a complimentary fabulous prize. How many guests sit through the pitch intending to win the free camera or toaster oven, but then end up spending 25K for a week at "The Falls" in Walla Walla?

But even in the time-share pressure cooker, no salesman is a match for an engineer. My wife and I sat there during the sales pitch until the inevitable question was asked: "What do you do for a living?" I would say, "I'm a civil engineer," and the salesman would inwardly curse between smiling, clenched teeth. The salesman would be thinking, "You're an engineer who has done all the precalculations, figuring out that this time-share resort deal is probably a financial sham. Even if it isn't, my pitch relies on shallow, emotional appeal and impulsive behavior, neither of which engineers are susceptible to. Why are you here wasting my time?"

Smiling, I would think back at the salesman, "I'm here to win the free prize, you idiot!"

So to the life insurance underwriter, I say, "Yes, I'm a civil engineer, darn it, and proud of it! Maybe I'm not that exciting. Maybe my clothes are functional but not that fashionable. Maybe I'd rather calculate than communicate. Maybe I'm not impulsive. Maybe I don't wear my emotions on my shirt sleeve, where they would interfere with the pocket protector. Maybe I'm dependable and dull. But at least, statistically, I'll stay alive longer than everyone else. I'm a good insurance risk!"

I can think of a lot more to discuss on this topic, but unfortunately, I'm running late. It's time for my bungee-cord jumping class. Today, we're going to practice on shear ice cliffs. I won't be wearing gloves.

The Twister

In my imagination, I can see formation of the twister. I'm standing in a field somewhere in Kansas. It's the Great Plains, and the field extends for miles in every direction. The horizon is distant and unreachable. It's a warm, muggy spring day. Cows are softly mooing in the field. The sun is shining, but dark storm clouds are building on that far horizon. This is the dry line, the leading edge of the storm. Flashes of lightning puncture the darkening sky. The cows look up from their grazing in concern. A massive black cloud forms, taller and more menacing than the others. The wind picks up, stirring dust in the humid stillness. From the bottom of the cloud forms a writhing cone that twists and turns and slams into the ground. There's a deafening roar as the twister begins its march across the flattened landscape.

I was asked to make a presentation at a Kansas transportation conference in mid-April. The request was sudden, offered two days before the conference began. For many people, the sudden request would have been a burden, but not for me. I've had a long-standing fascination with tornadoes. I've always wanted to see a tornado—a small one, from a safe distance, of course. In the northeastern United States, you don't see many tornadoes. But in Kansas, in mid-April, it's a different story. Spring is the prime time in Tornado Alley to see twisters.

My presentation was at Kansas State University, in Manhattan, Kansas, a small city about 100 miles west of Kansas City. It is possible to fly to Manhattan from my hometown of Boston, but you have to make at least one plane change, and the last flight is on a little propeller plane from Kansas City. On the day of the trip, a typical spring cold front extended across western Kansas. The high temperatures were in the 80s

(Fahrenheit) to the east of the front. To the west, a blizzard raged in Denver. This clash of hot and cold is the basic ingredient for development of tornadoes.

I was ready. On the first segment of my trip to Manhattan, the plane approached the Kansas City airport at night. As the jet plane started its descent, I was peacefully reading in my window seat. There were flashes of light outside beyond the wing. At first, I thought that the captain had turned on strobe landing lights, but on closer examination, that wasn't what it was. We were flying over an intense thunderstorm with continual flashes of lightning, one about every second or so. The flashes illuminated the roiling clouds below. I was grateful that the captain had found a calmer stretch of sky from which to land the plane.

From Kansas City, I caught the last leg of the trip on the little propeller plane. This flight reminded me of that scene in the Flintstones where Fred and Barney help the plane take off by running on ground below their seats. After a short, relatively smooth flight, the plane landed at the Manhattan Airport in the still of night. A gentleman at the airport counter told me that a big front with hail had just passed through an hour earlier, and another one was expected a few hours later. Apparently my plane had landed in a lull between storms, so I was batting zero for two. Maybe a twister was out there somewhere on the plains, but so far I had missed it.

Anyway, the gentleman told me that something about Manhattan, Kansas, resisted tornadoes. For no apparent reason, all the twisters avoided the valley and went on to blast Topeka or Kansas City instead. Still, this fact didn't stop the placement of warning signs on all the public buildings. For example, an ominous sign was placed at the entrance to the Kansas State Student Union. The sign had a picture of a black funnel and instructions on how to get to the basement in case of a tornado. When I arrived at my hotel, I turned on the Weather Channel for a few minutes before going to sleep. There were tornado warnings all over Kansas and Missouri.

Part of my fascination with tornadoes is related to the basic concept, which I have trouble accepting. How can it be that suddenly the atmosphere turns into this dark, malevolent force that can blow up cows and buildings? Other natural disasters are at least conceptually plausible. You can see the results of earthquakes but not the event itself, so there's nothing to visually debate, and hurricanes are essentially giant, windy rainstorms. But a tornado funnel, with its swirling concentration of evil and clear delineation between the good and the bad, doesn't seem likely or even possible. I needed to see one, in part, to confirm that it was possible.

Now that everyone has video cameras, we northeasterners can see plenty of pictures of tornadoes. Not far from where I was staying is the site of probably the most famous tornado–civil engineering video. In it, some TV news reporters are driving on the Kansas Turnpike near Wichita. Suddenly, a white tornado drops out of the sky. The white tornado starts chasing the reporters in their van. They race down the turnpike, finding refuge under a highway overpass (which, by the way, is strongly not considered a safe place to be). They walk up to the top of the abutment and crouch for protection from the wind between the steel–composite concrete deck stringers. Meanwhile, one brave cameraman is still running his video recorder. On the tape, you can see the tornado funnel approach the bridge. Then, the twister is on top of the bridge, in a rush of wind and dust. The last pictures from the video show the funnel's retreat on the other side of the bridge, and the aftermath: turned-over trucks and cars on the highway and a lot of scared bystanders.

At the beginning of my presentation to the transportation conference the next day, I asked the audience if it was really true that all the twisters missed the city of Manhattan. They laughed. With the presentation complete in the early afternoon, I packed up my stuff and prepared to go back home. The itinerary included three flights: first, a hop on the propeller plane from Manhattan to Kansas City, then a flight to Philadelphia, and finally back to Boston. I had no time to lose between each segment, with only 30 minutes for transfer at Kansas City. The propeller plane would have to take off and land on time for me to make it home.

Unfortunately for me, a great wind roared down the Great Plains that afternoon. On a clear day without a cloud in the sky, the sustained wind speed was 30 to 40 miles per hour, with faster wind gusts. The van taking me to the airport had trouble staying on the road. With a wind that strong, the propeller plane would have been blown clear back to Nebraska. I had only one chance to make it to Kansas City in time to catch the next flight—I would have to rent a car and drive the 100 miles to the KC airport.

I had a choice of cars. Being frugal, but stupid, I picked the small economy car rather than the big sedan. Without any time to spare, I raced out of the Manhattan airport parking lot and drove east on the plains of Kansas. I revved the engine. But with a fierce wind blowing across the interstate, I struggled to maintain the car in my lane, driving past swerving trucks and blowing livestock. The lightweight economy car shuddered and skidded in the wind.

It was a clear day with no clouds, but finally I could see it—the twister! I imagined that the twister was blowing just on the side of the road. It was a dark, photogenic, evil twister, like the one at the end of the movie, *Twister*. In this movie, actor Bill Paxton is driving the truck chasing the twister while all sorts of obstacles like combustible gas-tankers and houses are blown onto the road. Actress Helen Hunt screams at Bill, "Left!" and he has to swerve left to miss a tree, then "Right!" and he has to swerve right to miss tumbleweed.

I imagined that Helen Hunt was sitting there in the passenger seat of my car. She screams, "Left!" and I have to swerve left to miss a flying cow. She screams "Right!" and I have to swerve right to miss a trailer home. She screams "Full-Depth Pavement Reconstruction!" and I think, wait a second, that wasn't in the movie. The Kansas DOT was rebuilding the interstate roadway one side at a time. The construction staging had all traffic, eastbound and westbound, occupying one two-lane side of the highway while the other was dug up and rebuilt. So, in addition to the wind and my overwrought imagination, I contended with high speeds on a narrow road with bollards separating the traffic. Also, I wanted to see how they were building the job and what equipment they were using. All this had to be done at 65 miles per hour, because I couldn't slow down or I would miss my connecting flight. Tapping my ruby red slippers three times wouldn't get me home that evening.

Fortunately, I made it to the airport in one piece and in time to catch the flight. I remembered how in *Twister* Aunt Meg said to her niece, Jo (Helen Hunt) "Jo, you've been chasing these things all your life. It's what you do. Go, do it!" It's a rallying cry to go and chase things in life, in spite of adversity, a sort of personal anthem for me. I didn't actually get to see the twister, but I knew it was out there. I wondered: when a tornado touches down and no one is there to see it, does it really touch down? I read peacefully on my flight back east to Boston—no thunderstorms flashing below. But, out there, beyond my reading light, beyond the portal window and the airplane wing, somewhere on the darkened plains of Kansas grazes the twister.

Engineering Fashions

When I walked into the office the other day, a colleague casually said, "Nice tie." At first, I thought she was referring to the clever patterns and evocative geometry. Then I realized it was a comment about fashion, as in I was wearing a fashionable, nice-looking tie. I hadn't received a compliment about clothes before, so I wasn't sure what to make of it. Engineers have great abilities, but fashion usually isn't on the list. Scott Adams, creator of the cartoon *Dilbert,* has made millions with this observation, among others. For example, Dilbert's tie curls up, and this really happens when you keep the tie beyond its expiration date. You may ask, how do I know this? Let's just say I have empirical evidence.

Male engineers look for clothing that is functional and covers those areas that need not be exposed in public. Using this definition, a shirt can last many years even if it's frayed and stained. The concept of matching clothes is advanced and outside of the baseline definition. Many men, myself included, don't have the greatest sense of how to match clothes, and we rely on simplified, rule-based criteria. Consider the following example. In my case, there are two seasons: corduroys and khakis. During the cord season, everything must be gray or black, with red or maroon ties and appropriate solid sweaters. During the khaki season, all shirts are some shade of blue. At first, I only wore light-blue shirts, but then I became daring and got darker blue and navy. All ties for khaki season are blue-themed and reasonably match everything else. I know this because I asked someone who knew better and confirmed it. Rounding out the wardrobe are the two pairs of dress shoes, black for cord season and maroon for khaki season. Imelda Marcos, I'm not.

Such is my simplified fashion life. In the morning, I grab something very quickly and can't go wrong. Although I must admit that, one time,

I grabbed a right maroon shoe and a left black shoe and left the house that way. Wearing mismatched shoes is a fashion faux pas that should be avoided. Sometimes I think there should be the equivalent of Garanimals for engineers. Garanimals is a line of children's clothing where kids match tops and bottoms by picking the same animal—a bunny top goes with a bunny bottom. For engineers, we could have a system of matching engineering parts: the I-beam tie goes with the I-beam shirt, the concrete batch plant shoes go with the concrete batch plant pants, and so on. This would introduce some exciting variety into the engineer's wardrobe while still sticking with rule-based criteria that can be easily applied.

Change is purported to be good, but when it comes to clothing, I believe that change is not so good. This opinion again is based on the engineering baseline criteria for clothing, which assumes that unless there is an ice age or some other climatic shift, the old policies still apply. When Management invented something called "casual day," we engineers had to adapt. We received a memo:

> I am pleased to announce that we will have a business casual dress policy from Memorial Day through Labor Day. As a reminder, business casual attire means employees can dress casually in a manner that is acceptable in a professional services environment. This policy is at the election of the cost center managers, and they are responsible for its proper administration.
>
> To assist in implementing this policy the following guidelines apply:
>
> 1. Attire must be neat, clean, tasteful, and professional looking.
> 2. Acceptable business casual attire includes slacks, casual shoes and socks, collared shirts (long- and short-sleeved), and cotton chino pants.
> 3. Unacceptable attire includes, but is not limited to: T-shirts, tank tops, halter tops, denim jeans, jogging suits, sweat suits, spandex, beachwear of any kind, sneakers, flip-flops, and excessively revealing attire of any nature.
> 4. Modesty and discretion must be exercised at all times.

Notice right off the bat that the memo is written in engineer-friendly terms. That is, it features specific, clear criteria on what's what and what's not. I offer some commentary on the criteria: concerning item 3, who would wear spandex to the office, or for that matter, beachwear? Does this need to be pointed out? Are there some engineers who have spandex bathing suits, and have they considered wearing them to work? Fortunately (or unfortunately, depending on your perspective), this policy will nip the developing spandex beachwear fashion trend in the bud. Also, I note that on the list of "musts" in item 1, each is a descriptive rule that engineers can understand and apply. Clothing must be neat, clean, and so on. The requirement that attire must be "fashionable" is not listed, and it's just as well. Otherwise, who would understand it?

When Casual Friday was first announced on my project many years ago, we were in the design phase, and everyone dressed formally in suits and clothing that required dry cleaning. The concept of Casual Friday confused me, and I wasn't sure what to do. (During the project's design phase, my wardrobe featured two seasons of suits—dark gray for winter and tan/blue blazer for summer—and all shirts were white.) At first, I wore my business suit but took the tie off. That seemed casual, but looked pretty stupid. After a few months of exposure to casual dress, I was in blue jeans. Once the project moved to the construction phase and most people dressed pretty casually anyway, it was casual enough to not put on the tie.

Around this time, I rode the train with my computer programmer friend. He is an engineer's engineer's engineer, and he wore a T-shirt and jeans to work everyday. He was in a space that was pretty much beyond fashion, so my fashion dilemmas must have been pretty amusing to him. On non–Casual days when I fidgeted with my tie, he looked pretty comfortable in his scruffy attire. He won't end up on the red carpet at the Academy Awards, like that woman a few years ago wearing a green dress that was about to fall off. But with his focus elsewhere, he will develop some spectacular software and perhaps save humanity, and that's probably better.

Who Likes the Chocolate?

In Hershey, Pennsylvania, chocolate is embraced not just as a candy, but as a unifying theme for life. It's all a bit over the top. A group of students and I attended a leadership workshop sponsored by the American Society of Civil Engineers. At our hotel, the Hershey Lodge, chocolate was everywhere. Centerpieces on every table had chocolate. When you checked in, the hotel staff handed out chocolate bars. The rooms had complimentary packets of hot chocolate. When you went to work out (and in Hershey, Pennsylvania, you really needed to work out), posters of chocolate cartoon characters were posted cheerfully on the exercise room walls, the message being that once you finished on the treadmill, you were rewarded with another Krackel bar. You could go to the spa and experience something called a "chocolate wrap." Actually, the spa had several options involving chocolate and cocoa, whether it was to be eaten or bathed in.

Overall, chocolate is a very positive thing. A short Web animation asks "Who Likes the Chocolate?" (http://www.weebls-stuff.com/toons/47/). The answer is that everyone likes the chocolate. Driving into Hershey, we were a bit suspicious. What if the natives liked the chocolate too much? They might be obese, what with all that chocolate night and day. The region would have the highest per-capita use of acne medication—who likes the Clearasil? It was a Willy Wonkan explosion of chocolate.

It's refreshing and interesting to visit Hershey, because the town can be experienced as a real place and not another landscape trashed by sprawl. In spite of the worldwide trend for homogenization of place, the town has forged a strong identity and understanding of itself.

Hershey is the place for chocolate, and its infrastructure is developed and built around that rallying theme. There is the chocolate factory, which perhaps doesn't suffer much from NIMBY issues (after all, who doesn't like the chocolate?). You drive down roads named after chocolate. In the summer, a pleasing cocoa aroma is in the air. Hershey does suffer to an extent from construction of the anonymous, sprawling, glopscape that has overrun much of the rest of the nation. But the town's understanding of itself has resulted in a place that has not been completely overrun. An outsider's general impression is of a real place, not a strip mall/office park/housing development/parking lot. People come to visit Hershey because it is actually a place.

Is an infrastructure theme a good counterbalance to today's trends of dehumanizing sprawl, infrastructure homogenization, and loss of local identity? The chocolate theme may be a Band-Aid, Disneyland approach, a weak substitute for real planning. Besides, plenty of potential themes would not be as appreciated as chocolate and would not lead to as positive an understanding of place. For example, not too far down the road from Hershey is Harrisburg and the nearby Three Mile Island nuclear power plant. Like chocolate, electricity is generally considered to be a good thing. So, using electricity as a planning theme, the towns near the power plant could have street names like Electric Avenue, and Larry the Light Bulb could be the village icon.

Unfortunately, the rap for the nuclear power plant is not as positive as it is for the chocolate factory. Maybe people would appreciate the need for fewer streetlights at night (because everything would glow in the dark), but probably integrating and building infrastructure around a theme of nuclear power would not work as well as it does with chocolate.

Having a unifying theme, an infrastructure goal, might be a better approach to constructing infrastructure than what we do now. The way we build and develop infrastructure in the United States today, in general, is disorganized and poorly planned. We value private space, and ours is perhaps the best in the world. But public spaces are often treated as an afterthought. The private spaces are not designed and sited to create good public spaces. So we go from our luxurious, comfortable private dwellings to our Class A office buildings and spectacular shopping malls. In between, we drive around a dreary environment of trashed freeways and moonscape parking lots decorated by dying junipers. Driving and commuting can consume hours, because everything is inefficiently spread out. The idea of a more compact infrastructure that provides the same level of private luxury and service while also offering some of the same to the public realm is, in general, considered to be un-American. In places like Hershey, though, devotion to chocolate has overcome this judgment. In upholding the greater good, that of chocolate, the resulting construction is better thought-out and organized.

Driving home from Hershey, we imagined that there was a nearby village called Nestlé. The two villages had fierce rivalries. The high school football teams got into cocoa brawls during half time. There were battles with Hershey kisses flying in the air, and one could just imagine what might happen at the fondue parties. If a unifying theme is good for planning infrastructure, then peer pressure and rivalry would also be a benefit. But the rivalry between Hershey and Nestlé would be cordial and friendly—because everyone likes the chocolate.

The Maze

One way or another we often find ourselves waiting on line. Whether we are queuing for a ride at the amusement park, standing around to pay for groceries, or backing up at a traffic jam, we treat lines of various sorts as a tolerated nuisance. The queues require places for the lines to form, and infrastructure professionals are the ones who design and build the waiting facilities.

The implied promise of automobiles is the ability to go from Point A to Point B at any time, at the driver's convenience. This, along with the multiple vehicles in the garage and the chickens in the pot, is the essence of the American Dream. We strive for complete freedom of mobility. However, the reality is that getting from Point A to Point B is increasingly more like a nightmare than a dream. Congestion introduces wide variability in travel times. A 30-minute drive can take 30 minutes or two hours. So, at least along the most heavily traveled routes, the promise of unrestricted, unfettered travel is an illusion. The open road can be found in some empty, deserted stretches of Nevada, but that's about it.

Most people think traffic congestion is a bad thing, but some commentators have more positive things to say:

> Traffic is not primarily a problem, but rather the solution to our basic mobility problem, which is that too many people want to move at the same times each day. Why? Because efficient operation of both the economy and school systems requires that people work, go to school, and even run errands during about the same hours so they can interact with each other. That basic requirement cannot be altered without crippling our economy

> and society. The same problem exists in every major metropolitan area in the world.*

Therefore, congestion is good, because it reflects the activity of a healthy modern society. People are moving with some freedom in different directions because they have places they need to go.

I think there's some truth to this. For 20 years, I commuted by train. My work travel schedule was governed by when the trains ran. The commute was precisely coordinated based on which commuter train I would take in the morning and evening. When I became a professor at Tufts University, I started driving to work. This opened up a whole new level of flexibility and frustration that I had never dealt with before. For one thing, I could go to work and leave whenever I wanted to, and the multilane freeways beckoned with four shiny blacktop lanes all for me. There was never a risk of missing the train. But the flip side of this seeming freedom was the wide and wild variation in driving travel times. The train would deliver to and from like clockwork. There were maybe 10 times in 20 years when I was delayed more than 15 minutes. But in my new commute, driving times would vary by as much as 200%. For example, if I left the house by 5:30 a.m., I could typically make the 30-mile drive in 30 minutes or less. Leaving the house by 6:30 a.m. would add as much as 30 minutes or more to the commute, because the Southeast Expressway would be gummed up. No shiny freeway lanes beckoned me then. The open road was a sea of red taillights, with anaesthetized drivers listening to Howard Stern on the radio.

The vocabulary of response to traffic congestion includes relatively new entries such as high-occupancy vehicle lanes and traffic metering. One of the latest ideas is to have special occupancy lanes priced for congestion—you pay a higher toll during rush hour for the privilege of faster speeds. This ingenious engineering solution takes advantage of increasingly sophisticated toll technology. But it also butts against the American Way of free and equal travel for all. Why should the rich avoid traffic jams just because they can pay for it? HOV lanes in their original formulation required at least two or more passengers in the car. Passenger requirements are a more democratic approach—pursuit of privilege for all, provided you behave in a socially acceptable way to earn the privilege.

---

*Downs, Anthony. (2004). *Traffic: Why It's Getting Worse, What Governments Can Do,* Policy Brief 128, Brookings Institution, Washington, DC.

In the fall, my family has a tradition of visiting New Hampshire to view the changing leaves. Our autumn day has many stops, one of which is at a beautiful hilltop farm with an apple orchard and a maze. The maze is laid out on a cornfield, sort of like in the movie *Signs* but without the alien invasion. The maze has an entrance and exit, with confusing paths in between that you have to navigate. The maze is charming and somehow relaxing on a crisp, sunny autumn day. The cornstalks shimmer in the cold northwest breeze, and the winding paths lead you deeper and deeper into the field, with seemingly no way of returning. In a way, this is a ritualistic experience of waiting on line. It is satisfying at the end that we find the exit to the queue.

The Ronald Reagan Room

On a trip to the Transportation Research Board convention in Washington, D.C., I stayed at the Omni Shoreham, a beautiful, refurbished, older hotel. The Omni Shoreham is one of three hotels that host the annual TRB conference. The hotel is perched on a bluff overlooking Rock Creek Park. The refurbishment was truly excellent, but even so, I was greatly surprised by the quality of the room I ended up in for this trip.

I was assigned room number 325. It sounded ordinary enough. I grabbed my overnight bag (which doubles as a gym bag and was slightly inelegant for the likes of the Omni) and took the elevator up to the east wing. Based on the signs, I was directed to the end of a long corridor. I counted the numbers, but when it got to 325, the numbering seemed to stop. There was one last door at the end of the corridor for the "Ronald Reagan Suite." I looked around, and then I realized that the Ronald Reagan Suite was, in fact, Room 325. Was this a mistake? Did they assign me a multithousand dollar suite instead of a normal room? I tried the electronic room key, and sure enough, it worked. Room 325, or more accurately, Suite 325, was mine for the night.

Since I am perfectly comfortable lodging in a Motel 6, it took me about half an hour to digest the surroundings of Suite 325. I entered to a spacious foyer with a hardwood floor and a spread rug. There were vases on the cherry and oak tables. In fact, vases were all over Suite 325, and also artwork that didn't appear to be of Motel 6 quality. The foyer opened to a vast living room, which had a fireplace, a comfortable desk where I began typing this essay, and a beautiful patio overlooking the pool, grounds, and Rock Creek Park. This was surely one of the nicest patio views in all of downtown Washington, D.C. As a special bonus for civil engineers staying in this suite (not that there had been any before—I was the first), you could see the reinforced concrete arch bridge that carries Connecticut Avenue over Rock Creek. Adjoining the living room was a separate TV room (I think there were a total of four giant TVs in the suite). Also next to the living room was a spacious dining room, with table settings for ten, hardwood floor with a Persian rug, and an elegant chandelier. The dining room abutted a huge, full-service kitchen with lots of appliances and more vases.

When I retired to the bedroom, I found a room that was about three times the size of an average hotel room. The walk-in closet was closer in size to an average hotel room. The closet had lights that conveniently turned themselves on and off using motion sensors. I played with this feature for a few minutes to see how sensitive the sensors were. The luxurious bathroom also didn't disappoint. It had a separate bathtub and shower. I'd like to point out that the shower stall wasn't the usual cramped space, but more like a small room. If you lost the soap there, it could take a few minutes to find it.

So I stared and walked back and forth about 15 times to take in the sights. I called the front desk repeatedly to verify that I was, in fact, supposed to be in Suite 325. The convenient placard on the front door (one of two entrances to my suite, by the way), noted that the fare for this room was $2,000 per night, which was a bit more than my expense report would allow.

Now, fortunately, this story doesn't have a Cinderella-type ending. The clock didn't strike 12, and my room didn't shrink back into a normal hotel room. Also, I didn't turn out to be a guest on one of those cruel reality TV shows—you know, let's put up the civil engineer in the grand suite and then show his boss how he's whooping it up. The sun rose the next day, and I woke up still in Suite 325. I went to my meetings and presented my papers. Everything was about the same as it would have been. It turns out that the rich still put their pants on one leg at a time, but they stay in much nicer hotel rooms.

Before leaving to return to Boston, I checked with the concierge for the lowdown on how I ended up in the suite. One possibility was that the hotel had received a copy of my recent employee review, but this theory turned out to be inaccurate. The concierge explained that hotels tend to overbook, similar to airlines. They have a formula for calculating how many people are expected not to show up. This time, more people showed up than the formula predicted. At the moment I checked in, the hotel was out of regular rooms. But I had reserved my space, and the hotel was obligated to provide lodging. They could have sent me to a hotel down the street (Motel 6?), but it was actually cheaper for them to put me in the Ronald Reagan Suite, which was open for the night. Typically, the Ronald Reagan Suite was open a lot, because no one can actually afford to pay for it.

I suppose that, if I had known in advance, I could have planned a big party that evening for my engineering friends. I would have purchased cocktail wienies and champagne from the gourmet shop across the street and laid out platters in my fireplaced living room. But there was no way of knowing in advance, since it was a purely random event, and a big party with engineers is not all that exciting anyway. It was interesting to me that the whole thing was governed, in a sense, by the engineering process. The hotel used formulas for estimating flow and had to devise a workaround Plan B when the formulas didn't quite match flow to capacity. This time, the result of Plan B was a pleasant blip during my trip.

Bringing Out the Inner Civil Engineer

This is an essay about stopping to smell the roses, instead of trying to build the rose bush. First, to set the tone:

At 5:30 a.m., I climbed off my mountain bike on the shore of Leach Pond, about 10 miles from my house. I looked out to the east, where the sun had just started to rise over the horizon. A cool, smoky mist rose from the pond, blanketing the fir and pine trees on the far shore. The geese and ducks, who had been making quite a racket, quieted a bit. Maybe they understood the magnitude of the moment. As the sun made its appearance, the light changed. The red orb breached the tree line and illuminated the mist. For a few seconds, the world was awash in a reddish, glowing misty light.

OK, the tone has been set. I think civil engineering has a gleeful, childish side that we should participate in more often and embrace. This side of engineering gets beyond the calculations and the dry analytical nature of the work, and it helps us to appreciate the pure joy and wonder of what we're really doing. For me, being a civil engineer is akin to playing with my childhood block set or building sand castles on the beach. I never really got tired of doing these things but had to stop because of expectations and peer pressure. It is great to have children because now I have an excuse to do it again, and no one suspects that it's really me playing and not just being a good dad.

There is an emotional side of civil engineering that is related to the appreciation of the changes we have wrought. By training and personality, we engineers tend to be analytical and dispassionate. The process and calculations that we immerse ourselves in mask the reality of what we're really doing: building full-scale Tinkertoy models and molding the physical habitat of humankind. It is to our disadvantage to forget this reality, because, in a sense, what's really important is not the process but the product. Or, said differently, the process we become enmeshed in is such an abstraction that we forget the heroic nature of our work as we're working on it. To design and build infrastructure is, in fact, heroic. Every human being needs us to do our job and do it well.

The border between the emotional and analytical civil engineer has been well illustrated for me. In 2000, the Amtrak Acela high-speed train started running. This electric train races up the northeast corridor at speeds up to 150 mph. If you take the 5:05 p.m. commuter train from South Station in Boston, you can arrive at my home town of Sharon just in time for the Acela to zoom by on the adjacent track. You can stand on the platform, behind the yellow danger line, of course, and watch the futuristic, aerodynamic bullet train roar through the station. I know it's not as fast as some of the other bullet trains around the world, but it's still a pure rush of speed and adrenaline to watch it whoosh by.

One night, I watched the Acela pass with a colleague who had worked on a contract to evaluate the air plume and impacts to commuters on the station platforms. He described to me how the calculations predicted dust doing this or the airflow doing that. I didn't really hear him. All I could think about, or more accurately, all I could feel, was the sensation of the great roar of the train and the incredible sensation of speed as it passed. The calculations enabled it, whether it was the train passage itself or the protection of adjacent commuters. Yet the calculations were the process, and the whoosh was the product.

As engineers, how often do we forget about the product and focus on the process? The answer for most of us is most days, I suspect.

The Japanese Honshu–Shikoku Bridge Expressway Company has a terrific Web site (http://www.jb-honshi.co.jp/english/index.html). I've used this site for an introduction to civil engineering class that I teach at Tufts University. The company is responsible for maintaining three separate highway routes over Japan's inland sea. These routes feature a collection of some of the world's greatest bridges, including the Akashi-Kaikyo suspension bridge, the world's longest span bridge, and the spectacular Tatara cable-stayed bridge. What's really amazing about the bridges is not the individual achievements of the different structures, but the incredible assembly of so many bridges in a line. Overhead photos show a lineup of multiple cable-stayed spans, followed by suspension bridges, with trusses, arches, and everything in between. These routes are a bridge-builder's dream run amok, a giant Tinkertoy set brought to life. In class, I pretend to be a wise professor, expounding on abutments, cables, tension, and bolts as I surf the site. But what I'm really doing is opening a photographic cookie jar. I hope that the students realize this and get the message.

Maybe we can come up with practical applications for the expression of this civil engineering glee at upcoming meetings. Let's build sand castles on the beach! Let's play with blocks! It's time to get back in touch with our inner civil engineer.

The Baby Sitter-In-Law

I am fortunate to be married to a woman much friendlier than me. Lauren has a personality the opposite of mine, which is closer to the cliché of the introverted engineer. She has made many friends over the years and maintained contact with many of them going back decades. She used to baby-sit for the kids across the street. Then the kids grew up. We were invited to the boy's wedding in Tarrytown, New York. My wife remembered all of the family members and friends, and they all remembered her. As part of the wedding's social structure, she was introduced as the groom's baby sitter. That made me the baby sitter-in-law.

The wedding was set at an elegant conference center/hotel, perched on a hillside about a mile from the Tappan Zee Bridge. The setting featured an outdoor ceremony on a beautiful patio overlooking the Tappan Zee, the three-mile-wide estuary of the Hudson River north of Manhattan. The Tappan Zee Bridge dominates the site. The bridge was built in the mid-1950s. It has several miles of low trestles, which rise to a hulking, cantilever truss main span across the shipping channel. As part of the New York State Thruway, the bridge carries routes I-87 and I-287 over the river.

I have a copy of an old promotional film that documents construction of the bridge. The video starts with a scene of two puppets riding a seesaw. This illustrates the concept of cantilever construction. The accompanying music sings about the seesaw, while a male 1950s-type announcer expounds upon the marvels of construction. The Tappan Zee Bridge was a monument in its day, and a trendsetter in many ways. At more than three miles, the structure was one of the longest over-water bridges built and to this day still ranks pretty high on the list. The cantilever span, while graceless, dominates the Tappan Zee. Even in that wide expanse of water, the massive bridge makes a visual impact, in the same way the Forth Rail Bridge does over the Firth of Forth near Edinburgh, Scotland.

The promotional film includes many dated scenes, such as builders tossing red-hot rivets through the air, violating just about every safety rule today. At this point, the film makes a comparison to baseball pitchers: that's one hot fastball. The film was set in the days when the popular view held that men built bridges and women cooked, so there are many politically incorrect scenes extolling the virtues of the male bridge builders. My favorite scene is when the announcer exclaims in his beefy voice that, "They're not just men. They're *bridge* men!"

On the other hand, the construction sequence pictures and descriptions are still interesting and educational. At the bridge site, the Hudson River widens to its estuary. Below the river bed is a layer of soft marine clay hundreds of feet thick. The design relies on pontoon action of hollow caissons to reduce vertical loads on the deep foundations. The cantilever section was staged from both piers off falsework on the back spans. The center closure members were rotated and locked into place, an impressive operation.

Not much has changed on the Tappan Zee Bridge since its opening. The shared breakdown lane in the center median has been converted to a seventh lane. A movable barrier system has been added to change the center lane's direction from eastbound to westbound and back, depending on the time of day and traffic demands. An automated transponder reader has been installed at the toll booths for those who buy a pass. This has eased congestion a bit, but the bridge often suffers from monumental traffic jams. Discussion has it that something needs to be done to replace the bridge or expand capacity. This will become one of the great transportation projects of the near future. In some ways, the challenges are more daunting now than when the bridge was first built. The east and west banks of the river are more densely settled than they were in the 1950s, leading to tougher construction staging and permitting. Environmental requirements for dealing with hazardous materials have become much more stringent, and these must be dealt with when building in the polluted muck in the river bed.

The Tappan Zee Bridge exemplifies the highway building period in the United States after World War II. Its construction permitted rapid development of the suburbs north of New York City. The bridge and connecting highways allowed for a new type of transportation and infrastructure development. The ramifications of this sprawling development pattern are just now being seriously debated. On a smaller scale, the bridge allowed widely dispersed guests to meet, somewhat conveniently depending on traffic, for the wedding party in Tarrytown. The unstated assumption is that the bridge will provide easy and convenient access over what would otherwise be a near-insurmountable barrier to transportation. This assumption is true for all of our transportation facilities, which quietly provide the infrastructure that is the basis of the way we live today.

So there I was, sipping champagne on that patio overlooking the Hudson River. You could hear the faint hum from the cars on the Tappan Zee Bridge. It blended in with the keyboard and soprano saxophone music. During the cocktail hour, all of the relatives and friends wandered around, as they will at a wedding. When some of them came over to say hello, I introduced myself, and they said, "Oh, you're the baby sitter-in-law."

Reston Town Center

Reston Town Center is a place you'll probably be hearing more about in the years to come.

With the American Society of Civil Engineers headquarters located in Reston, Virginia, many committee meetings are held there. The town is conveniently located a few miles from Dulles Airport, and it's easy to fly in and out. For a recent meeting, we stayed at the hotel in Reston Town Center. This visit was a bit different than other visits to Reston.

Reston Town Center appears to be a traditional, small town center. It has a main street next to the hotel where we stayed, with shops and pleasantly scaled sidewalks. Trees line the sidewalks. There is a big fountain and a center square. You can go to a fancy coffee and pastry shop and sip cappuccino sitting at tables on the street. The street has a movie theater on the end, with a traditional marquee and lobby entrance. The architecture here is deliberately non–shopping mall. Everything is laid out in urban blocks. At the plaza with its beautiful fountain, you can dine al fresco at one of many restaurants.

Reston Town Center is a lively village. It evokes thoughts of American town centers of yesteryear. On Saturday nights, teenagers hang out in the center, going to the movies and visiting the ice cream and malt shop. Families with children stroll the streets. During the work day, office buildings off the center streets are filled with workers, who go strolling at lunch time.

As a place, Reston Town Center is a bit deceiving.

For starters, the center is not really a town center at all, not in the sense of an old, established town crossroads that developed over decades. Reston Town Center is a multiuse planned development. The streets, hotel, office buildings, shops, and movie theaters were laid out and designed all at once by architects and engineers. What's different is that instead of the typical strip mall/shopping plaza/office park, with minimal landscaping and acres of parking lots, the developers consciously attempted to recreate a town center that Reston never really had. The resulting development is perhaps one of the few places in northern Virginia where you can sip cappuccino at an outdoor table and stroll along a pleasant, human-scaled streetscape.

To understand and better appreciate this development, let's compare it to the surrounding area. Northern Virginia around Dulles Airport is perhaps Ground Zero for demonstrating the problems of sprawl. This region has been built up almost completely within the last 20 or 30 years. As suburban sprawl goes, the area is relatively well planned. There are wide, well-designed freeways and wide feeder boulevards with state-of-the-art intersections. Most construction follows in the pattern of American suburbs, with commercial areas separated from office parks separated from different types of housing. The housing developments themselves are carefully separated by the zoning: luxury single family, not as luxurious single family, town houses, and so on. The design quality of the new infrastructure in northern Virginia is top-notch, grade-A work.

At Reston Town Center, you can do things you cannot do elsewhere in most of northern Virginia. When you leave the hotel, you can walk out on the street and shop, stroll, go to the movies, go to a meeting in the adjacent offices. You don't have to drive there. It's interesting to look at the border between the town center and the surrounding turf. It turns out that the development is ringed by parking garages. Neighboring suburbanites can drive their cars to Reston Town Center, park in the garages, and then stroll and enjoy urbane street life. In this way, Reston Town Center is more like a Disneyland street than a real town center. The development itself appears to be very successful, but the impact on development of the surrounding areas is mixed. The streetscape is being expanded a few blocks with new office towers. On the other hand, further away are the more typical housing developments and strip malls. Housing is one thing that has not been integrated into this mixed use development. When you log onto the Reston Town Center Web site (unlike many town centers, this one has its own web page, http://www.

restontowncenter.com/), luxury housing is advertised. But the housing is not really part of the center, because it's separated by wide, traffic-clogged arterial boulevards. So, there is an uneasy boundary between Reston Town Center and the surrounding suburbia. The development appears to be successful, but it has not transformed Northern Virginia into a version of itself. It is more like an island of urbanity in the suburban sprawl sea.

All of this can be important to us because the issue of urban and suburban sprawl is poking its head at the top of infrastructure agendas. Civil engineers are now directly or indirectly dealing with these issues as part of environmental reviews and development plans in many of our infrastructure projects. As we plan and design our traditional works of transportation projects and infrastructure development, we see that the development rules are different and changing fast. Successful project work requires a greater multidisciplinary understanding and appreciation of the new rules. At Reston Town Center, you can see many of these infrastructure issues illustrated, full scale. A visit can offer a good lesson about the contrasts between "smart growth" design concepts and sprawl.

Don't Throw This Away, I

Being a pack rat is a common personality trait among engineers. You can go to almost any engineering office and see piles and piles of saved stuff. There are boxes of old calculations, mounds of design drawings, copies of reports going back to the days of George Washington. We engineers never know when we're going to need something, so it's important to save it. In triplicate. We are well stocked should a disaster strike. We may not have food or water, but at least we'll have plenty of old project documents to look over in the dark.

The need to save things is ingrained in our training. We are taught from the first day of engineering school that we must start with a strong foundation. At the beginning of structural design class, we learn the equation and bending curve for a simple beam. The next day, the beam is not so simple. It becomes complex. But we had better save the references and equations for the simple beam, because the two-span continuous beam uses the same starting formulas. So, we engineers learn that every piece of information is built on some other piece of information. Just to be sure, we had better save everything.

I think a natural selection process is going on here. It's not just that learning to be an engineer trains us to be pack rats. It's that the natural pack-rat personalities gravitate toward engineering. I remember taking those open book examinations in school and then later the P.E. exam. It's good to be prepared for a test, but many engineers go way beyond what is required. For the P.E. test, some candidates arrive with multiple boxes of references. If permitted, they probably would drive in with a forklift and bookcases on which to arrange their texts. In an eight-hour exam, the smartest engineer can't possibly consult that many references.

However, I don't think that is the full story. It's more about engineering peer pressure, keeping up with the Joneses, the intimidation factor. The one who has the most references wins.

Of course, being a pack rat has its drawbacks. We can't save everything. There's a point when we have to throw stuff away, with the associated feelings of loss and regret. One of the more traumatic moments in an engineer's life is moving to a new home or office. This is when the rubber meets the road. The engineer is faced with sorting through years of accumulated things and deciding what to save and what to trash. A lot of the debris is junk, of course, but each piece of paper, each report, each obscure text, has some engineering emotional value. Parting is not easy.

Some of us look to computers to be our salvation. In the near future, we will live in a paperless engineering society, with drawings and reports all electronically stored. In the present, however, we have twice as much paper and now a new source of clutter in stuffed hard drives and overflowing e-mail accounts.

Keeping all of this in mind, I decided to confront my failure. I resolved to use my excellent engineering analytical skills to identify the problem and solve it. I am a pack rat by nature, but I will choose not to live like one. I will throw out the lecture notes from high school, the interim conceptual design submittals that may have some relation to some work that I may do some day. I will fill up the recycle bin with old junk and achieve a spotless desk. I will turn over a new, uncluttered leaf.

Then, I got a phone call. A colleague remembered a report that I worked on five years ago. It was, I thought, an obscure report on a topic that I hadn't considered much since. But the caller thought the report was relevant to a project he was working on. He wanted a copy. I scoured what remained of my piles of reports. I was able to reconstruct parts of the report from old word-processing files (saved on my over-full hard drive), but I think I had thrown away the printed copies in my efforts to be organized.

So saving everything wasn't such a bad idea after all. You never know when it will be needed. The next day, I ordered a new bookcase.

Don't Throw This Away, II

Somebody famous once said that we all get at least 15 minutes of fame. My 15 minutes started when I wrote "Don't Throw This Away" as a column for *PB Network,* the Parsons Brinckerhoff corporate technical journal. An editor of *Engineering News Record* read the column and proposed that a version of the essay appear on the *ENR* opinion page. I was told that *ENR* has a larger reader circulation than *PB Network* (by about two orders of magnitude). This is said not to disparage *PB Network,* but to illustrate the appeal of the proposal to my ego.

The editor wanted some artwork for the page. An illustration was prepared showing the top of a hairless engineer's head behind mounds of paper at his desk. They also included a photo of me (still with hair).

The essay was printed in *ENR's* November 1998 issue, and it generated a buzz. People faxed me copies from across the country, with the words, "YES!! THIS IS ME!!" printed on the transmittal sheet. Engineers greeted me in Boston, sheepishly promising that they would immediately go back and clean up their offices. Everyone seemed to identify with the essay, although not everyone got the point. For example, one colleague routed a copy of the essay throughout his office and then saved it in the clip file.

Ironically, as I worked on the rewrite for *ENR,* I had the opportunity to live through one of the defining experiences described in the essay. After 10 years, it was time to pack up my office and move from South Station to our Kneeland Street office a few blocks away. Here I was, writing about the trauma and tribulations experienced by a hypothetical engineer during an office move, and suddenly it wasn't so hypothetical. It was reality. It was me combing through piles of important junk, mounds of back project documents, and dusty conceptual design reports. It was me filling multiple dumpsters while agonizing over every tossed report that might rescue a future project. It was me confronting my pack rat self, and it wasn't a pretty sight.

During this excruciating period of packing and moving, it occurred to me that I missed something in the original essay. As I excavated the old documents, I went back in time. I visited myself at an earlier age, during the time when the junk was originally saved. I wondered about the motivations of my younger self in accumulating all that paper. What caused the youthful me to save all those old reports, papers, and product literature?

I concluded that one motivation was engineering insecurity. Younger engineers feel the need to demonstrate their knowledge, to prove themselves to their elders. They are unsure of their professional roles and contributions, so they save everything, thinking that the documents will bolster their status and input.

It takes many years of experience to understand that what is most valuable for an engineer is not saving the information, but knowing how to use it. Once you realize this, you can become comfortable with the idea that you don't have to have everything at your fingertips. Engineers who learn this lesson sooner can become organized sooner. It need not take filling five dumpsters on moving day to figure it out.

I have some advice for younger engineers, many of whom are starting a career of clutter. You can lead an organized, stack-free life. You don't have to be a pack rat. Follow these guidelines:

- If it sits buried in a pile for more than a week, it's useless. Throw it out.
- If you save it out of a feeling of insecurity, with the thought that you had better save it or else, it's useless. Throw it out.
- If you don't have the space or time to manage the documents you save, such as periodicals and project papers, the documents are useless. Throw them out.

- A neat, organized office is a greater virtue than the feeling that you've saved something somewhere for some use sometime in the future. Throw the stuff out.
- Everything written by me should be saved. You never know when it will be needed.

I'm looking forward to another 15 minutes.

Don't Throw This Away, III

The collapse of Enron and the subsequent scandal produced the specter of key documents being shredded and tossed. Enron's accountant was accused of disposing of all sorts of sensitive documents before their time. This was reported to those of us outside the scandal as being on the scale of the Iran–Contra affair or the erased Nixon Watergate tapes. To avoid the potential for a similar catastrophe, we received a visit from the Clutter Consultant.* The Consultant's job was to instruct us on what to file and how to file it. This professional was a master of the new discipline of saving things. We were requested, via written memo, to attend Clutter Class.†

We all sat around a conference table, armed with notepads and copies of documents. The Consultant began with a discussion of the purpose of the class. Overheads detailed the Enron debacle and the sensitive nature of information. We had handouts (which we were supposed to save) that taught us in greater detail what we were supposed to save. It was a fairly thick handout, but the executive summary was that we were supposed to save everything.

For example, suppose the contractor sent a Request for Information (RFI). This would be a written question to the design engineer filled out on a form, something like "Your design has egregious mistakes, leading to loss of millions of dollars and months of delay. More specifically, should the #4 rebar on drawing X really be #5? Please advise, and do it fast." The RFI form would be sent with an attached, written transmittal form to the Resident Engineer. The Resident Engineer would maintain a file of all RFIs and responses. We designers would receive copies of

*This was not the actual title of the consultant.
†This was not the actual title of the class.

the RFIs to prepare the response. In the past, there was one file of RFIs. However, according to the Clutter Consultant, if we wrote notes or had draft calculations in preparation for the response, this draft information had to be filed and saved. Draft information, even thoughts, showing the progression to resolution, was to be considered pertinent data that should be saved. We had to order a whole new set of file cabinets to catalog what used to be draft, working information.

The need to save essentially everything conflicts with aspirations to declutter. As natural pack rats, we engineers work in offices that often have that post-seismic-event appearance. Once, in my office suite, the safety department determined that the mounds of papers and documents all over the place—on the floor, in the corridors—were becoming a hazard. Something had to be done. The safety department issued photos of the top offender offices as negative examples to the rest of us to clean up… or else. This was sort of like the FBI's list in post offices of the 10 most wanted fugitives. At the top of the list was an engineer I'll call "Bob." Bob is a terrific structural engineer, very thorough, smart, and patient, and one of the easiest people to work with. But Bob does have a slight problem with clutter. When his office appeared on the most cluttered list, he took action. He threw out cabinets-worth of documents. He threw away an old sandwich, and he tossed the orange on his shelf that had desiccated with time and turned into a citrus mummy. It took him about a day and a half, and it was stressful for him. But it was worth it! Bob ended up with an organized, uncluttered office, at least for a week. Unfortunately, not long after this dramatic transformation, the flow of paper caught up and surpassed the capacity of the wastebasket. Things have pretty much returned to status quo in Bob's office.

I aspire to the ideal of having one piece of paper on my desk. It is a theoretical ideal, the file cabinet version of Utopia. I want to save nothing, but the prevailing trends seem to lead elsewhere. I taught a class on bridge engineering at Tufts University for graduate students. The students had the notion that I was an all-knowing structural engineering presence, able to answer every question. I'm not sure where they got this notion from. Well, that's not really true—I carefully cultivated the notion. I lectured to this class from a pretty tall pedestal. Three of my students were junior practicing engineers studying at night for their master's degrees. When not preparing for my class, they were busily working on a structural inspection of bridges north of Boston. The bridges included rebuilt railroad structures on the northern commuter rail line: some prestressed slab bridges and a single-span, steel through-girder bridge. The three junior engineers wanted to know who designed the railroad bridges, which were rebuilt as part of the Commuter Rail Improvement project in the mid-1980s. Having the drawings and calcs would make their inspection job easier to do. They asked me, since I knew everything.

It turned out that the one who designed the bridges was me. I looked through a stack on one of my dusty bookshelves and, sure enough, found the original calculations. Someone else had saved the drawings, so I was able to get copies of those as well. The students were suitably impressed: not only did I know everything, but I had designed every bridge as well.

So between the Clutter Consultant and my own experience as a professor, the message seems to be that everything has to be saved. At least, until the next Most-Cluttered list is circulated.

Don't Throw This Away, IV

A while ago, when I was working at Parsons Brinckerhoff, it was announced that our e-mail system was a dinosaur that had to be replaced. The IT people concluded that the prehistoric software was essentially running on fumes, with official support from the developer long since dried up. The system constantly crashed. Every other day or so, messages sent to the office in Boston would end up in Saskatchewan.

In the original plan, the new e-mail system would directly replace the old one with no hassle. You would log in one morning and all of your addresses, mailing lists, and other doodads would have been automatically transferred. This sounded like a good plan to me, since I pretty much live and die by e-mail. I couldn't imagine life without automatic lists and instant, multiple communication paths. Also, without functioning e-mail, I would have to develop a personality and speak to people on the phone and things like that.

Unfortunately, the organized planned transition turned out to be not so organized. As the fateful date approached, the entire computer system failed, piece by piece. We had hit the iceberg, and it was time to person the lifeboats. I fidgeted with the computer settings, desperately seeking ways to transfer all of my comfortably automated data to the new system before the information was lost. Unfortunately, the old data had been accumulated for years, and it needed to be decluttered. So once again, I was living the saga of the cluttered engineer, with this episode coming in a hi-tech, electronic form.

The first step to reach the new e-mail nirvana was to send a message to everyone using my old e-mail list, because apparently the new system would generate an address book out of this. In this massive mailing,

I caveated and apologized, saying that the message should be ignored and deleted. Of course, I got back dozens of responses. Some people, to whom I hadn't spoken in a while, e-mailed back to say hello. Some thoughtfully provided additional comments. "Linda" wrote: "And here I thought you were just being friendly and sending along some good gossip. What a disappointment."

"Lisa" sent a helpful message updating me on the status of several people from her company on my list. Some people had left. Some went to Florida. Some had died and no longer needed e-mail addresses.

Eventually I was up and running in the new system. Slowly but surely, I learned the nooks and crannies and had the beast tamed. But then, electronic clutter reared its ugly head again. Our system change had taken place at about the time the e-mail world was experiencing an exponential increase in spam. Every day, my shiny new e-mail account received dozens of these unsolicited messages. Now you should understand that I am a bit insecure to begin with, so receiving multiple messages on how to improve on one's limited natural endowment wasn't helpful. Also, I got a lot of messages for discount Viagra, discount mortgages, and mortgage brokers that offered discount Viagra if you signed up for mortgages.

This would have been funny, but it was widely reported in the news media that e-mail spam was a serious problem. Spam caused billions of dollars of lost time and revenue, and it overloaded the computer systems and servers. The overload probably contributed to the collapse of our old system. The underlying pathology of e-mail spam was similar to other forms of clutter. That is, there was no obvious penalty to the ones creating the clutter. In fact, much the opposite was true. The more spam advertisements that were sent out, the better, because the e-mail spam ads were essentially free. So what if only 0.001% of the recipients actually bought the product and improved their endowment? That was free money to the product hawkers, who paid no penalty for the electronic clutter they caused.

Spam blocking software is available, and spam has become a game. My new system has a slot to block all messages from the spam addresses, but apparently the senders change a letter or two and are in business again. Parsons Brinckerhoff added software to block messages from porn sites. Under this approach, the software searches for naughty words and blocks the message. I sent one message with what would have been a naughty word in a different context. The recipient, "Gordon," responded to me and repeated the word, also in a perfectly acceptable, not-naughty context. So his response had the naughty word written twice. Apparently two naughty words was enough to trip the software, because a chastising message came back. Probably a message was sent to Gordon's Permanent File as well.

Eventually, the new e-mail computers were up to speed, and we all lived through the experience. Service was spotty for a while. Whenever the system crashed, I would go to Bob's office. We would toss around his new desiccated orange and mull over memos written in 1988. It was like a blackout, sort of. I thought, this must be how the people lived before e-mail. It was a slower, less informationally intense time, and there were fewer cc's.

An Ideal Geotechnical World

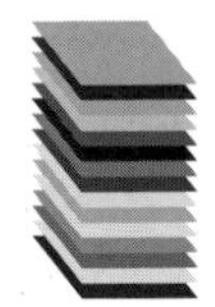

I am envisioning an ideal geotechnical world. In this ideal world, simple geotechnical exploration leads to an idealized, but completely accurate subsurface characterization. The exploration is such that the engineer has a complete understanding of the type of soil and rock, almost down to the soil molecules. Understanding of the soil characterization is further enhanced by simple, easily performed, accurate tests that yield precise information about the material behavior.

The data are easily digitized and manipulated and fed into complex but robust soil–structure interaction models. The models feature time-dependent, three-dimensional, construction-staged, inelastic, completely accurate representations of soil behavior under any conceivable type of loading and stress path.

Geotechnical engineers have learned to harness this incredible analytic technology to make accurate and reliable predictions for foundation strength, construction soil movement, stability, and other calculations. The analysis is done by sophisticated software with great ease—so simple in fact, that it could almost be done by a structural engineer.* The rigor of the analysis is not orders of magnitude more accurate than its input data. Analysis results are not multiplied by a factor of safety of 16.

In this ideal geotechnical world, I attend meetings with the contractor. Dialogue from the meeting goes something like this.

Me: "Contractor, your contract specification includes language to identify unknown obstructions that we think may be there, but we can't be absolutely sure until you start digging."

Contractor: "We knew there'd be obstructions. No charge."

*I'm kidding. Structural engineers are as smart as geotechnical engineers.

Risk is shared fairly in the ideal geotechnical world between owner, designer, and contractor. However, due to the exceptional advances in analysis, design, exploration, and construction, there is no risk, so sharing it is less of an issue.

Unfortunately, it turns out that soil is not a very ideal engineering material. Dealing with soil is sometimes closer to dealing with people than, say, the comforting but boring regularity of steel. Soil behaves in ways that are often difficult to quantify and predict. Another problem is that soil is found in the ground, where you can't really see it. The best you can do is scoop up a few samples and poke at them, trying to learn their secrets and extrapolating resulting behavior. These are the realities dealing with geotechnical analysis and design. The best engineering practice is to recognize and appreciate these realities. Two words should come to mind when working with soil: "judgment" and "reasonable". Engineering judgment is applied to determine conclusions based on results of analysis and exploration. The resulting design and construction application needs to be reasonable—the results should be understood and applied in the overall context of the problem.

Advancing technology and understanding will lead to improvement. But the truth is that soil is soil, and progress is likely to be incremental. For now, the unfortunate truth is that we don't live in an ideal geotechnical world.

Infrastructure and Coming of Age

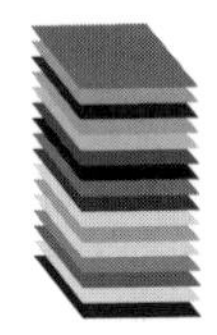

Our understanding and impressions about place and infrastructure are formed when we are very young, and they are colored by our life experiences. This seems obvious, but it's not something most of us think about, even the professionals responsible for designing and building the infrastructure. These impressions also weigh heavily on our attitudes about what infrastructure is good or bad and influence our reactions to engineering projects.

When we are young, the environment that we grow up in becomes what we expect the environment to be—it's our baseline experience. The built form of this environment includes houses, cities, highways, and all other components of infrastructure, as it is situated in the natural terrain and landforms. Our view of how the world should look physically is characterized by this first set of impressions. These views can be drastically different for different people.

For example, citizens of the United States, perhaps the most culturally homogeneous of large countries, live in regions that physically are very different. I was raised in the Northeast. In this part of North America, rainfall is greater than the worldwide average. There is a lot of surface water in rivers, lakes, and reservoirs. Cities and towns are designed around bodies of water. So my unspoken understanding of how a landscape should look includes scattered ponds and rivers peacefully flowing in the background.

When I travel to the Southwest, I'm startled by its dryness. There is little rain, and water doesn't just lie around in lakes as it should. Lakes do exist in this landscape, but typically require a dam such as the one at Glen Canyon in Arizona. The resulting body of water, Lake Powell, doesn't look like a lake at all, but it appears as part of a weird, beautiful, alien landscape. The wide open deserts and ranges of Arizona must seem perfectly natural to the people who live there. A child growing up there must expect to see wide expanses of rock and sand. In contrast, the Northeast would seem strangely closed-in and soggy to a southwestern child.

Reaction to new infrastructure depends upon who is reacting. A northeasterner planning a large facility in the Southwest may believe that the land is essentially empty and ready to be filled. Southwesterners may react much differently, appreciating the natural and dry open spaces that shouldn't be marred and closed in. On the other hand, westerners may have no problem with filling in a marsh, since to them a marsh is a bizarre, useless landscape.

The passage of time further complicates the way that an individual perceives the built environment. The infrastructure that seemed so natural to us as children becomes what we rebel against as adolescents, although we may not think of it that way. The comfortable (I hope) surroundings of home, which are handed to us by our parents, suddenly look boring and wrong. A new frontier is exciting and calls to us.

I remember my father being quoted in our hometown newspaper. I don't remember the topic of the article, but his quotation was about Orangetown, the town that I grew up in. He talked about the time when he first moved to the town and how it was a beautiful, growing place. I couldn't understand what he was talking about. The same place, through my eyes, was boring and not very attractive—just a standard suburb with the usual infrastructure accoutrements. The quotation stuck with me, and as I got older I better understood what he was talking about. The beautiful place that he saw was through his young adult eyes, since he had just moved from the city to start a family. For me, Orangetown wasn't the frontier, but the boring, expected place where I was born. In other words, my father and I saw the same place but understood it much differently.

This psychology is at play in Boston. Hundreds of thousands of Americans, myself included, migrated to Boston for college. These Americans will always see the city through a youthful haze, even when they're older. The type of infrastructure as represented by Boston will always have certain connotations to this group. Old, dowdy, red-brick

Boston for this group was where they ran away from home and became young adults. So, for example, the giant neon Citgo sign in Kenmore Square, which would be an ugly electronic billboard elsewhere, is for many folks a beloved, artistic beacon and historic landmark. For countless students, it lit up the way home from late-night beer parties.

This same psychology can have practical implications. On an infrastructure project, it's important to understand the attitudes of the people to be affected by the project. These attitudes will be influenced by their baseline understanding of infrastructure, that is, their set of (unspoken) expectations. A large number of Bostonians, native and newcomers, will prefer to see Boston the way it was, with few dramatic changes. This is why the Boston Red Sox will probably play in Fenway Park for a long time to come. The places where we spend our salad days will always remind us of being young, even when the places themselves are old and stodgy.

Learning the Expanding Body of Knowledge

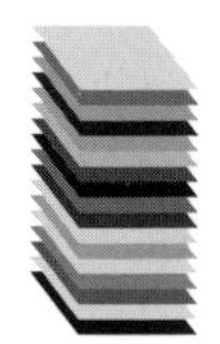

How much knowledge is enough? Once the bite was taken from the apple, a lot more apples kept growing on the trees. An apple a day keeps the doctor away, but a required consumption of bushels and bushels of apples leads to indigestion or worse. The American Society of Civil Engineers admirably took up the task of defining the civil engineering Body of Knowledge, or BOK.* The BOK is a detailed outline of everything a practicing civil engineer should know. But if the BOK keeps on growing, the effort to define it, much less learn it, becomes a moving target. It's the old problem of walking halfway to the wall, but never reaching it. Although in this case, the wall moves away as we approach it.

The BOK continues to expand exponentially. At the same time, proposals for revising engineering education suggest that training should include more focus on the "soft" knowledge skills, such as communication, management, writing, and leadership. This leads to a paradox. Engineers are expected to be technically proficient, while at the same time be skillful communicators and leaders. All of this 10 pounds of material is supposed to fit in the same old five-pound bag or even a four-pound bag, as state budgets redefine university course loads and as available education hours decrease.

It should go without saying that the first requirement for an engineer is to be skilled in engineering. Someone has to know how to do it. Everyone's quality of life depends on good, functioning infrastructure.

*In this essay, "BOK" refers to the overall collection of knowledge. Some also refer to BOK as the process and committee work involved in evaluating the related issues. The BOK, as in the overall knowledge, is described in this essay as expanding exponentially in an unorganized fashion (which is what knowledge does). However, the committee focusing on these issues is well organized.

Not a day goes by without this implicit understanding being met. The bridge is expected to not collapse when you drive on it. The water must make it from the well or reservoir to your spigot. The traffic lights must switch from red to green in some reasonable, rational interval and not turn blue or fall off the pedestals. Although fruit may grow by itself on trees, infrastructure functions don't happen by themselves. They are enabled by skilled civil engineers. Because the skilled civil engineers are mortal, new skilled engineers are needed eventually to replace them. This is achieved, we hope, by education.

However, it is not enough for engineers to be technically proficient. Engineers must be able to communicate about their work, function in teams, and lead others who are not proficient in the work. Failure to master nontechnical skills results in engineers who perhaps know what they're doing but have no impact and no say in the events around them. So without the "soft" skills, the "hard" skills can be of limited use.

The BOK continues to grow, but there is a limit to how much can be learned. We can deduce this by applying the Law of Learning Equilibrium. A student is alive for only so many years. We don't know exactly how many years, but we can use statistical averages to illustrate the concept. During the student's lifespan, time must be subtracted for periods not dedicated to education, such as sleeping, infancy, watching movies, and other nonacademic pursuits. After other adjustments, the remaining time is the maximum amount available for learning. In an episode of the old *Star Trek* TV show, Dr. McCoy placed an alien mind device on his head and instantly learned how to perform complex brain surgery on Mr. Spock. But using the reasonable assumptions that this technology will not soon be available and that the rate of learning has an upper limit, we can deduce that there is also a limit on the size of the BOK that can be absorbed during the available learning period.

The implied limit has several practical results. If the BOK continues to grow and only so much can be learned, then at some point parts of the BOK must be trimmed so that the new stuff can fit in. Some knowledge must be determined to be no longer relevant or important. Consider, for example, slide rules. A few decades ago, slide rules were considered an indispensable tool for civil engineers. But now the knowledge of slide-rule theory and application is rightfully considered antiquated and no longer part of the engineering curriculum. An organized, in-depth review of the overall civil engineering curriculum would probably identify many subjects that could be judged obsolete. The curriculum could be streamlined or revised to free up time for new knowledge, which is being discovered as you read this. But no one has sat down and tried to quantify explicitly what should stay and what should go.

Trimming the BOK tree is something that happens anyway, but it generally happens by default, without an organized process. Many of us would consider this (if we considered it at all) a result of the marketplace of ideas. In the marketplace, those ideas such as slide rules that are no longer relevant are discarded from active consideration. The formal education process, however, is less like capitalism and more like communism. Instead of an idea marketplace, the professor, or textbook writer, or both decide what ideas are relevant and present them as the syllabus. Students are taught until graduation that some things are supposed to be known, and here they are—no marketplace, just centralized planning. The shock comes after graduation, when the budding engineers are thrust into the marketplace to intellectually fend for themselves. The education process is then called "gaining experience," an informal method where an apprentice engineer, with any luck, will have guidance from good mentors to show him or her the ropes.

The ASCE BOK committee has taken on the daunting task of trying to reel all of this in—to define the BOK (at least in outline form) and to propose how and when it should be learned. The challenges include not just the moving target, but the lack of clear consensus among civil engineers on what device should be used to hit the target. Should a master's degree be required? Is the BOK to be achieved by practical experience? How is practical experience measured? Can the BOK committee goals be achieved in the United States if engineering is increasingly outsourced offshore?

While the questions are tough and the process is hard to define, to paraphrase the movie *Apollo 13,* "Failure is not an option." We will continue to drive around expecting the bridges to stay up and the traffic lights not to turn blue. We will be in big trouble if those expectations are not met.

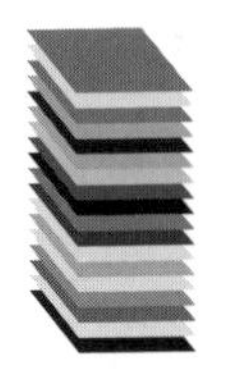

The Road Not Built

It is disconcerting to new graduates that much of what they design in the engineering office will never be used. In school, students work on neatly defined problem sets and projects. Their resulting work is the answer, what is expected. The students submit their work, the job is considered complete, and then the next assignment is tackled.

Unfortunately, outside of the neatly defined environment of school is the mess of the real world. Many projects are designed to varying degrees of completion but never built. File cabinets may be stuffed with calculations, but the physical landscape is not cluttered by these imaginary structures. These projects become virtual bridges, buildings, and tunnels, supported by painstaking calculations and details, described by volumes of documents. But they exist only in the imagination of the engineers. Robert Frost's poem claims that the road less traveled made all of the difference. I've often wondered about the road not built.

The Internet age has brought an explosion of communication and a corresponding increase in the documentation of everything. Now we can surf Web sites describing projects that were never built. The University of California at Berkeley set up a library project describing the bridges around San Francisco Bay (http://www.lib.berkeley.edu/news_events/exhibits/bridge/intro.html). One section, "Unbuilt Projects," includes documents showing bay crossings that never came into being, as well as discarded project designs of bridges that eventually were built. We think of the Golden Gate Bridge with reverence and awe. It is certainly one of the most beautiful bridges ever built, at one of the most spectacular locations. But early schemes for the crossing, as documented by the Web site, depict much different structures. One early scheme was for a hulking, monstrous combination cantilever truss–suspension bridge, which surely would have been one of humankind's ugliest structures had it made it off the paper. Frank Lloyd Wright had a vision for a second Bay Bridge crossing. His sketches show a low-slung, utilitarian concrete bridge, unlike the iconic, grand double-suspension crossing that had already been constructed. Other schemes for crossing the Bay had artificial islands and villages, all very Californian.

On the other coast, the New York metropolitan area has seen many schemes for projects that never made it to construction. Some of these projects are included on a Web site dedicated to New York crossings (http://www.nycroads.com/crossings/unbuilt/). Where there is a body of water without a crossing, someone thought about bridging it. For example, the eastern end of Long Island was to have a 30-mile-long series of causeways and suspended spans connecting Orient Point to New London, Connecticut. Real construction had proceeded to the point that part of a highway was built, connecting the end of the Long Island Expressway at Riverhead to the new bridge, but there's no new bridge at the end, only a ferry. The highway dead ends on the north fork peninsula of the island.

Several bridges were planned to cross Long Island Sound. Probably the most notable, and the one that came closest to construction, was a bridge to be built from Rye to Oyster Bay on Long Island. This bridge was planned as an extension of the Cross Westchester Expressway, part of I-287, the New York metropolitan area's beltway. Another freeway, the Seaford–Oyster Bay Expressway, was built on the Long Island side to receive the traffic. The bridge plan had I-287, which currently terminates at I-95 in Rye, continuing south up and over the Sound to Long Island. The proposed bridge required approaches through somewhat wealthier and more influential neighborhoods in Long Island and Westchester County. The project engendered fierce opposition from residents along both shores who didn't want their backyards and pristine frontage on Long Island Sound spoiled by highways. The result was one of the earlier successful highway project rebellions in the late 1960s, and as a result, the bridge was never built. Now the engineering plans grow moldy in storage.

But politics is not only to blame. The engineering process itself results in a lot of design effort with nothing constructed to show for it. For U.S. highway projects, the Federal Highway Administration requires two options, one in steel and one in concrete. Both are designed to 100% completion, but there is at least a 50% chance that one design will never turn into a real bridge (or more than 50% if the project is cancelled and nothing is built). Even without the choice between steel and concrete, engineering by its nature is iterative, so for every beam hanging in the air, there are hundreds of imaginary beams from the earlier design schemes.

We engineers take this process for granted. We all know that every time we put pencil to paper, the results will need a lot of fine-tuning and redesign. This is natural for us and expected. But, by accepting this iterative process, maybe we unwittingly buy into attitudes that devalue engineering work itself. For example, if a bridge design requires two options, there is at least a 50% chance that one of the designs will never be built. It is then easy to consider the work behind the unbuilt bridge option as not real design work. From there, it's a short step to viewing the work required to prepare the unbuilt option as so unimportant and trivial that it can just be thrown away. Nontechnical parts of the process—the politics, the publicity—don't feature the same degree of iteration. The public relations people don't work on the "steel and concrete" political alternatives and completely throw one away. If some of the technical engineering work is of such a value that it can just be shelved, then maybe it's not so important. Or at least that's the way it can seem working on the job.

We engineers probably won't have much success in building the roads not built. The nature of our work mandates iteration, change, and designs that never leave the CAD screen. But one thing we can do is celebrate the good work that is done, regardless of what is built. If the steel option is selected, then we can still be proud of the concrete option and document it. The conceptual design process offers a treasure trove of engineering thinking and development, no matter which scheme was chosen. In the dust bin of unbuilt bridge designs lies amazing engineering work, and also, perhaps, some opportunity for lessons to be learned from the less amazing work. Web sites such as the ones at Berkeley and

New York show how we can celebrate and increase the value of what we do. It would be good to feature more of this as part of the process—to not just do the work, but to appreciate it, learn from it, and celebrate it. To do so would be good for education and good for engineering. It would also help for teaching nonengineers about what we do, how we do it, how hard it is to do well, and how important it is to understand and support the process to get the best results.

Build It and They Will Come

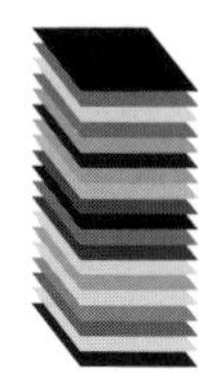

Debate over highway construction in the United States has been divided into two camps. The pro-highway camp believes that if we build enough roads and freeways, traffic congestion will become uncongested. This simple flow-versus-capacity argument is obvious to the proponents of the first camp, but mocked by the members of the second camp. The anti-highway group believes in a version of the movie, *Field of Dreams,* in which the main character hears a voice that says, "Build it and they will come." Members of this group intuitively understand the obvious conclusion that the more freeways we build, the more traffic they attract—build it and they will drive on it. By this argument, it is impossible to build our way out of congestion, because the new roads just lead to more congestion. So why bother?

At the extreme, the two sides have no love lost for each other. The pro-highway builders consider themselves rationalists, unlike their fern-loving, sprout-eating, transit-riding opponents. The anti-highway crowd understands that the paving and sprawling over of the countryside is not progress, but a march toward environmental Armageddon. For them, it's not only about highways, but also about all sorts of other seemingly peripheral issues like wealth redistribution and obesity (because people drive two blocks to get a quart of milk instead of walking, which is more exercise). These extreme arguments and related opponent-bashing, while perhaps entertaining, don't focus well on the problems and how to solve them. There are kernels of truth from both sides. A paper by Robert Cervero in the *Journal of the American Planning Association** creatively sheds light on the discussion. Cervero throws a bone

*Cervero, Robert. (2003). "Road Expansion, Urban Growth and Induced Travel." *J. Amer. Planning Assoc.* 69(2), 145–163.

to both the pro- and anti-highway supporters and then proceeds to beat them on the head with it.

Cervero attempts to test objectively the theory behind "build it and they will come." His detailed analysis and evaluation of results of 15 highway projects showed that building new highways does not necessarily attract more traffic and cause more congestion. It's true, and somewhat obvious, that highways built in semirural areas at the urban/suburban edges attract sprawling development, especially at the interchanges. This spaced-out development leads to more vehicle trips but does not necessarily cause more congestion. Cervero cites the Houston area as an example of a place that has been able to build enough freeways to reduce congestion. This approach seems possible: if we pave over enough, we get enough additional capacity regardless of the number of new McDonald's and Wal-Marts at the exit ramps. But, as Cervero notes, even with congestion reduced, "...residents of places that are able to build themselves out of traffic congestion might not necessarily like what they get." The issue, then, is really about land use. Cervero writes:

> It is important to focus on the bigger picture when framing highway policies. The problems people associate with roads—congestion, air pollution and the like—are not the fault of road investments per se. These problems stem mainly from the unborne externalities of the use of roads, new and old alike. They also stem from the absence of thoughtful land use planning and growth management around new interchanges and along newly expanded highways. (Cervero 2003, p. 160)

The issues are being played out in real-time all across the United States. Every metropolitan area is spreading out. Consider the case of North Carolina's largest city, Charlotte. This formerly quiet southern town in the last few decades has burst at the seams. In terms of highways, it had been served by two interstates (85 and 77), with a little loop freeway downtown and a short east–west freeway connecting to the airport. With the city expanding outward, the existing freeways needed to be rebuilt and widened. NCDOT has been building one of its most ambitious highway projects, a new beltway around the entire metropolitan area. Sections of this new expressway, I-485, have opened to the south and east. This is a spectacularly well-designed freeway, with excellent geometrics and clean, attractive bridges and structures. In terms of road design, what has been built so far is probably at the pinnacle of highway practice—a superb job.

So why isn't everyone in Charlotte pleased? Probably because the specter of what happened to Atlanta haunts the city. Atlanta is rated number one, or near number one, in traffic congestion and sprawl. The 20-lane freeway to the airport still has traffic jams. Residents of Charlotte are afraid that their city will become like Atlanta, and they have good reason to worry. The new freeway interchanges in Charlotte have quickly been paved over and built up. The I-485 interchange at Providence Road, for example, used to be wooded and rural. With the newly opened freeway, it features a clash of arterial boulevards, with a new shopping center, middle school, housing pods, and office lagoons. All of this is well-designed, modern, upper-end infrastructure, but in the aggregate, it generates thousands of car trips. Nothing can be done, gotten to, or accomplished without going by car. This is by design.

The stakes in this debate are high and directly relevant for those working in the A/E world. The worst-case scenario is paralysis—no work, no projects, and unsolved problems that will get worse with time. To avoid paralysis, we engineers have to become better at identifying the problems and focusing our analytical skills on solving them. Unfortunately, blaming all the problems on roads and autos masks the real issue, that of land use. But in the United States, land is private property, and individual ownership is a right not easily tampered with or debated in the public realm.

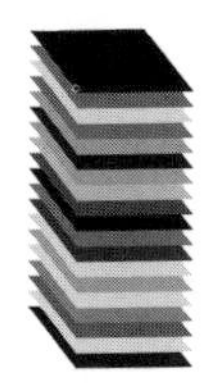

New Car

A car is an inanimate object. It is a machine. On a dark and stormy night in July, I drove my old station wagon down the driveway and out to the freeway. This had been our family car for almost six years. It used to be comfortably cluttered, with the kids' school papers and candy wrappers spread over the floor. But for the first time since we bought the car, it was spotless and clean. The only reminder of clutter was the smell of ice cream dropped on the back seat. For the wagon, this was to be our last ride together.

When Lauren got a new job farther from home, I did the usual engineering evaluation and determined that it made sense to get a new car. The old wagon was reliable and comfortable and probably could have stayed in service a few more years. But I wanted the relative reliability of a new car so Lauren wouldn't have to worry. We shopped around, enjoying that pleasurable experience of interacting with car dealerships. We weren't desperate and could bide our time for the best deal. Yet it was interesting how, even in that situation, the dealers got under our skin. Their presentations featured an impressive collection of feints, double-talk, and marketing misrepresentation. Each of their special clearance sales began with an offer for thousands of dollars off a price that had been raised by thousands of dollars.

We did our research and waited for the right moment. As these things will happen, suddenly the right moment arrived. Despite my engineering nature and need to weigh and evaluate everything, we made a quick deal. We purchased a station wagon almost identical to the old one. It was a slightly darker shade of tan, but otherwise, it was hard to tell the cars apart.

Two days later, I was going to pick up the new vehicle. En route to the dealership out on Interstate 95, my old, comfortable car handled well and without complaint. But I wondered—did it know? I had just finished reading Eric Schlosser's book, *Fast Food Nation.** In one chapter, Schlosser describes how cows turn into burger patties. He is visiting the slaughterhouse. Outside in cramped pens, the cattle are fattened with offal, road kill, and other animal remains. When it is time for "processing," the animals are directed to a chute. The chute is winding and deceiving so the cattle can't see what and who is waiting at the end. One cow seemed to know better and wouldn't go down the chute. It bellowed and protested, looking right at Schlosser with sad cow eyes. Probably its name was Elsie. After some coaxing, the creature was redirected to meet its fate. Elsie sadly looked at the author one last time, bellowed, and not long after that someone enjoyed a Quarter Pounder with Cheese.

I thought about a car advertisement I had seen on TV. The ad is told from the viewpoint of an old car (this assumed that cars had viewpoints). It is sitting in a junk crusher, about to be flattened for scrap metal. But just as the claws bear down on the vehicle, the scene changes, and we see happy moments from the car's past. The owners are driving the car in sunlight. Kids are playing and spilling ice cream on the back seat. In the end, we see the owners driving a new, similar car. The implication is that the spirit of the old car has been reborn in the new vehicle that the owners have purchased. Apparently, car spirits never die. They just become the subject of new car loans.

Cars are inanimate machines, but we attach all sorts of anthropomorphisms and emotions to them. In the United States, our infrastructure is designed around automobiles. Many people don't go a single day without riding in a car. I've spent much of my professional life designing structures for use by cars: bridges, roadway structures, and highway tunnels. However, I had never thought about the car as being much more than a car.

When it was time to leave the house for the last ride, my wife became very emotional. She had driven the old station wagon for many years. The children were young in this car. We were young in this car. We had gone on vacation in this car. At the moment when she would last see the vehicle, she was teary-eyed. My daughter crawled into the back

---

*Schlosser, Eric. (2001). *Fast Food Nation*, Houghton-Mifflin, New York.

seat and didn't want to get out. It didn't really matter that the new car we bought was almost identical to the old car, right down to the color. The old car was taupe. The new car was Arizona Tan.

I made it to the dealership just as the thunderstorm intensified. The skies darkened and opened up. Sheets of water poured from the heavens. I pulled the old wagon into a parking spot for the last time and went inside to fill out the paperwork for the new car. The sale had been concluded, so there was no need for more wheeling and dealing. The auto dealers were a bit beside themselves. Wheeling and dealing was what they did and not being able to do so was unnatural. After a few minutes of banter, I was handed the new keys. I walked out into the pouring rain and greeted my new station wagon. Somewhere in the distance, the old wagon shivered and bemoaned its fate. Was it waiting for me to come back and get it? I never returned.

Soon I was driving on the freeway. The heavy rain poured down on my new windshield. Because of the darkness of the storm, the gloom of night had prematurely descended. I breathed deeply of the new car smell. On the crisp, full-sound stereo system, with substantial woofers, Dan Fogelberg was playing. It was the song where he meets his old lover in the grocery store. They had an encounter in the frozen foods section. In the end, Dan and his old lover part. He turns to make his way back home, and the snow turns into rain. For me, it was July, and the rain started out as rain. But as I turned to make my way back home in my new Arizona Tan station wagon, the saxophones wailed that "Same Old Lang Syne."

Acronyms and the Explosion of Useless Data (AEUD)

I attended an engineering project meeting where I discovered that we had invented our own language. We were sitting there merrily chatting away in initials and all sorts of verbal shorthand. We talked about C11A1 and RFIs, whether it should be a DR or an NCR in the B3. A guest at the meeting from outside the project tried to follow along. Soon his eyes glazed over. However, we knew what we were talking about. We had created a special club with our own language, and the guest wasn't a member. It reminded me of a scene from the movie, *2001, A Space Odyssey,* which used to be set in the future. Now that it is the future, this movie is an appropriate reference for everything in life. In this scene from the movie, we are back at the dawn of time. A tall black slab has landed on the African plains, and a bunch of gorillas are jumping up and down in front of it, gurgling and screeching. This must have been what the meeting seemed like to our guest. He was lucky that the *2001* analogy stopped there, because in the next scene, the gorillas pick up antelope bones and beat the visitor to a pulp.

The technical complexity of life now requires abbreviations for us to communicate. It takes too long to talk about everything in longhand, so we have resorted to creation of a verbal shorthand. In the English language, some shorthand becomes commonly accepted in the vernacular. We all know what a VCR is. Fortunately, this device stuck around long enough so that we can call it by its proper initials. Unfortunately, now the VCRs are being replaced by DVDs, and soon the VCRs will be like slide rules (SRs). I will have to store the videotapes (Walt Disney: The Entire Collection) with my set of LPs.

The explosion of information is outstripping my ability to remember things. They changed the telephone area code at my mother's house. I grew up with this telephone number. It worked well for decades, so I'm not sure why they had to change it. I needed to dial the number the other day, and I couldn't remember the new area code. I had to use Information. When you have to dial 411 to call your mother's house, you know there are problems.

There's a cause behind all of this. As usual, it's the computer's fault. The common denominator behind the new language and the exponential increase of area code numbers is the vast explosion of information enabled by computers. Human beings can't collectively remember, store, and process all of this information, but computers can. The problem seems to be that there is no obvious penalty for storing and processing useless data (GIGO, or for those who need things spelled out: garbage in, garbage out). In civil engineering, and I believe in all aspects of life, we are losing our ability to weed out the junk, because it seems like we don't have to. In the dark days before computers, we couldn't remember and deal with so much data, so we didn't. Anything that didn't fit in wasn't fit in, a kind of natural selection for useless information. For example, the original Central Artery Project in Boston included design and construction of the green viaduct that has now been demolished. This original project had a set of drawings and reports that was maybe 1⁄100 the size of the newer incarnation. True, CE projects are much more complex today than 50 years ago, but by a factor of 100? The reality is that if you have a word processor, you can write endless reports and documents, and we do. Then, you need acronyms to shorten the time it takes to talk about them, and that's how you end up with a new language.

A degree of self-discipline is necessary to weed out the useless information before it becomes information. Because data storage and management are so much easier, on such a greater scale than ever before, we don't do much weeding. Maybe that big computer company can program this into the software. In addition to the green squiggly line for bad grammar and the red squiggly line for bad spelling, they could add a purple squiggly line for something stupid that shouldn't have been typed in the first place. Like spell-check and grammar-check, this would be intelligence-check. If the user violated the program rules too many times, the keyboard could deliver an electric shock to reinforce the point. Someone would have to define the rules on what qualified as being stupid, but this hasn't been a problem for that big computer company before.

So here we are in the 21st century, playing an evolutionary catch-up game with the machines and data we've created. This is not exactly a new story, but the pace seems to be accelerating and the problem seems to be getting worse. Most significantly for me, I'm getting tired of acronyms (IGTA).

Postscript I

A colleague from another university was working on a project to evaluate structural parameters for analysis. The idea is that you can back-figure appropriate parameters for a particular structural model based on the actual structure's response to static and dynamic loading. Then you can place measurement instruments on structures such as buildings and bridges. By using the strain and movement data, you can compare measured results to predictions from the analytical structural model and determine over time the condition of the structure. The basic idea is simple, but implementation is complex. There are limitations on how much you can measure, how accurate the data are, how well and accurately you can mathematically define a structure, and other problems. The approach is still largely the subject of research and not yet a widely used practical tool in the field.

In order to determine the parameters, you need a pretty complicated computer system to evaluate the data and perform parameter estimation. My colleague developed a system, and she named it:

Parameter **E**stimation **M**odel **U**pdating **S**ystem.

Although it was very complicated, the system seemed to be doing a good job, and my colleague was very proud of its success. I wasn't at the meetings where the system was evaluated and discussed, but I imagine that the conversation went something like this:

My Colleague: We are here to discuss the Parameter Estimation Model Updating System.

Professor A: Thank you for inviting us here for this discussion. We have evaluated your Parameter Estimation Model Updating System and it seems to be doing a good job. Many people are satisfied with your Parameter Estimation Model Updating System.

My Colleague: This is good to know. I'm pleased that people are satisfied.

Professor B: Yes, it seems to be quite effective. Is it a large system?

My Colleague: Yes, my Parameter Estimation Model Updating System is quite large.

Professor A: That's interesting, because some studies have indicated that size does, in fact matter, for a Parameter Estimation Model Updating System.

Professor B: Although there has been significant discussion in the field that the system's size is not necessarily as important as its method of application.

Professor A: Yes, I agree with Professor B.

My Colleague: I would think that the system's size is not important as long as it gets the job done.

Professor A: An excellent point! Efficacy matters more than size. But even so, do you have plans for growth of your Parameter Estimation Model Updating System?

My Colleague: Yes. I plan to add several new features to the system.

Professor A: Can the Parameter Estimation Model Updating System do a good job at evaluating structural member stiffness?

My Colleague: Yes, estimation of moment of inertia is a key feature of the system.

Professor B: So, you have a robust Parameter Estimation Model Updating System?

Professor A: A system that is poised for growth and will function better in the future?

My Colleague (perhaps a little perplexed by now at the discussion): Yes, one could say that.

Postscript II

The name of the computer system has been changed.

The Quacking Moment

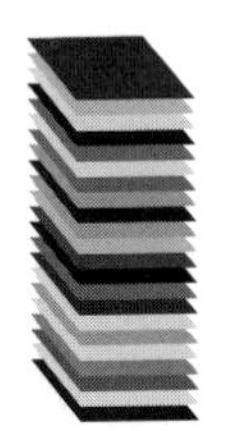

The other day as I was lecturing to my concrete design class, I started to imitate Elmer Fudd. I was trying to describe the "cracking moment," the value of bending moment on a concrete beam when the first cracks occur. But instead of saying "cracking moment," I pronounced it as "quacking moment." Instantly the class burst into laughter, and valuable teaching minutes were wasted. I put on a stern expression and tried to regain the high, serious ground, but it was a lost cause. Even I couldn't keep a straight face, and I was the professor. Unfortunately, the pedestal I stood on couldn't handle the load from the quacking moment.

I'm not one to miss the opportunity for teaching a good lesson, so the next day I developed a special PowerPoint presentation. This contained a brief discussion of the background and theory behind quacking moment. If the weight of the duck is taken as P_{duck}, then on a simply supported beam of length, L, the quacking moment, M_q, is equal to $(P_{duck} \times L)/4$.* I also provided a helpful illustration (Figure 1).

My brush with the quacking moment illustrates a larger trend affecting the communication of technical issues. With the increasing complexity of everything, spoken English and technology have been at odds for a while, and things don't seem to be getting better. I remember one of my college professors who provided an early, negative contribution to my engineering education. This professor was lecturing about the "soap film" analogy for stress, in which the shape of a soap bubble can help illustrate magnitude of stress. But he pronounced "soap film" as "soup film." I had no idea what a "soup film" was. I had this picture in my

*This is funny for structural engineers. If you're not a structural engineer, note that the maximum bending moment for simply supported beam of length L, with load P placed at the center of the beam span, is calculated as $PL/4$. Now that you know, it's funny for you, too.

Figure 1. The Quacking Moment

mind of dirty tomato soup, with a scummy film on top. For years I was clueless about the "soup film" analogy until I finally realized that "soup" in this case was "soap."

I take pity on this poor teacher of my past, because however bad his pronunciation was, mine is not much better. I suspect that my own students are not greatly better off when I mangle phrases during a lecture. Instead of wondering about soup films, they will have to deal with the "quacking moment" every time they design in concrete. They'll be working in consulting engineering offices in the near future. The chief engineer will ask them to design a concrete beam or footing. Everything will be OK, but then for no apparent reason they'll burst into laughter. The chief engineer, who is probably not the most secure chap to begin with, will look at his tie or discreetly check his breath. He is not aware of the impact of the quacking moment.

But even if I should become a better enunciator, able to consistently pronounce complex words, my problems would not be over. New and increasingly bizarre words are being invented every day, and they are being combined in complex, incomprehensible phrases. Attempts to simplify with abbreviations and verbal shorthand often make things worse—in addition to the long-winded technical terms are the endless abbreviations. Whole subsets of language contain information of essential importance that can only be understood and spoken of by a small group of experts.

Added to this Babellian mix is the trend of fading of American-centrism. (You will note two invented terms in that sentence.) Maybe some are pleased that after decades of U.S. dominance in the post–World War II period, other cultures are advancing in terms of worldwide influence. But in the past, English was the de facto business language of the world, and as other economies catch up, other languages rightfully take their places in the spotlight. This means that in addition to struggling through technological miscommunication in English, we will have to be proficient in the mispronunciation of Chinese, Spanish, and other languages as well. It is a worrisome thing—adaptation to the comprehension of exponentially expanding knowledge doesn't seem to be keeping pace with the rate of expansion. Reread the last sentence, and you can see what I mean.

But wait, there's more! The increasing development and refinement of complexity has led to niche languages and groups that speak them. Here at the university, you can attend lectures where a handful of people know what's being talked about. Everyone else nods and smiles as if they did. But even if and when the ideas are approachable, the jargon is completely lost on all except a very small, "in" crowd. A computer program is available to help write bogus technical papers. The program scans your documents, extracts random words, and assembles sentences that appear to have nouns, verbs, adjectives, and other components of English. To a reader, the sentences seem to be sentences that appear to be combined in paragraphs. Yet the entire creation is gobbledygook. It's totally funny to read, especially at those times when the passages really seem like actual, technically correct gobbledygook.

Assuming that the governing theory of human development is that of evolution and not "intelligent design," at issue here is the exponential, sudden explosion of human knowledge and the terminology needed to communicate it. The theory of evolution posits the concept of natural selection, where genetic changes gradually occurred over millions of years, favoring adaptations that improved a species' survival. With a long enough period of time and a stable enough environment, the generations would gradually improve and evolve. So, subtle physical and mental adaptations would favor those premodern humans who were able to escape the saber-toothed tigers and live to procreate. But how can humans evolve to meet the new environmental challenges of information overload? I remember when there were no PCs or word processors, and I'm not that old. There aren't yet enough descendents of me to succeed or fail at this new environmental challenge—no natural selection. I'm still here, I'm trying to adapt, and I actually give lectures about things to other human beings. So far I'm succeeding, because I haven't been eaten by a saber-toothed tiger, but I have survived to pass on my inability to speak to future generations. So this negative trait has not been unselected by natural selection, at least not yet.

In the overall geologic timeline, this is all very sudden. After millions of years of life on earth, human technology has exploded in a few short decades. Like everything else, issues related to the explosion of knowledge and our inability to communicate it will probably (and hopefully) sort themselves out. These are turbulent times, and the fittest and best pronunciators will presumably attract mates, procreate, and live to lecture another day. Meanwhile, my poor students will have to learn to adapt. They won't have to work through the soup film analogy, but they will need to account for the quacking moment.

Fish

The Pike Place Market in Seattle has a fish store that is more than a fish store. It is a cottage industry.

The fish store starts out looking like a lot of other fish stores. This shop is at the head of the market at the junction of Pike Street and Pike Place. The market itself perches on a bluff overlooking Elliot Bay and the Alaskan Way Viaduct. The market includes an esoteric collection of shops and vegetable stands, some open-air and some enclosed. It is a bustling, popular tourist destination, and the bustle is greatest at the fish market.

When you walk up to the counter, a performance is going on. In a typical fish store, attendants passively wait for you to order flounder, and then they chop it up and stick it in a bag. But at this fish market, the counter is alive with flying fish. The shopkeepers are busy flinging salmon, perch, and other aquatic items through air, at customers, and into bags. If you order a pound of salmon, five attendants call out loudly and in unison, "Salmon!" and fling the salmon. In between the fish flinging, the attendants go out into the crowd, which is really an audience, and entertain with jokes and fish stories.

According to the lore of the fish store, this all started innocently enough. The shopkeepers wanted to generate some business and activity at their store. It helped that they were all either naturally ebullient or a bit over-caffeinated. With all of the fanfare, performance, and flying fish, soon the fish really did start to fly, and not just from the hands of the shopkeepers but out the door with pleased customers. Sales ballooned, and the store got a buzz as a place to visit. The store owners were onto something. They weren't content to sell fish at this point, but they decided to institutionalize their achievement. They sold videos on how to motivate people. The videos became very popular in project management seminars. The owners' message was simple, somewhat obvious, and something many of us forget: any job can be done better, with enthusiasm and heart. Any effort can be better. It's 99% attitude and 1% something else, but what does the 1% matter if you've got the 99% down?

Every effort we make can be thought of as a simple, personal choice: do it really well, or slog through it. I am not well known as one of those chirpy optimists, but after a while I decided that the simple conclusion taught by the fish store was true. Each person really does have a choice. The lesson is perhaps of particular importance to engineers, because, at least from what I've seen, we tend to be dour and introspective and forget about this basic truth. For example, if you're designing a concrete beam, and you select #6s at 12, you can choose to marvel at the theory that makes this simple calculation correct and even life-saving. You can be amazed at the millennia of experience and theory, of pozzolan concrete and Mount Vesuvius, of sparkling concrete cable-stayed bridge pylons rising to the Ohio/Kentucky sun. This may sound totally stupid, but in reality it's a better way to do everything. If you don't believe this, then go visit the fish store!

And that was my assignment. My wife Lauren, a human resources manager, was teaching a series of classes on motivation. She purchased the fish videotapes and wanted me to take photos of the store when I visited Seattle. So I went to the market, almost got clocked by a piece of flying cod, and returned with the photos. As part of her classes, she bought motivational, stuffed fish. The red one was named "Pete Fish" and the yellow one was named "Pete Fish." Apparently they were both named "Pete Fish." There was a problem. Lauren wrote:

> On a cold, yet sunny day in February, the unthinkable happened. One of our two "stuffed animal" fish was "fishnapped" from his cozy home in the conference room. Our beloved red Pete Fish was replaced with a Ransom Note. If we did not meet the fishnappers' demands of catnip and whole milk, which were to be left near the door by Press A, our fish's goose would be cooked! Unfortunately the ransom note did not state

on which date the ransom was to be left, so we were unable to meet the requirements. Fortunately, the following week red Pete was returned safely to Human Resources, but not before a sadistic fishnapper had "grilled" him a bit.

You would think all's well that ends well, but the day before red Pete's return, our yellow Pete was fishnapped. I am still waiting for the ransom demand. If you have any information as to the whereabouts of our yellow Pete please contact your HR representative. A reward is highly likely for yellow Pete's safe return!"

Now, in keeping with the lesson of "Fish," I should make something out of all of this. After reading a bunch of these essays, by now you may have noticed that typically a wistful, clever observation ties everything together.

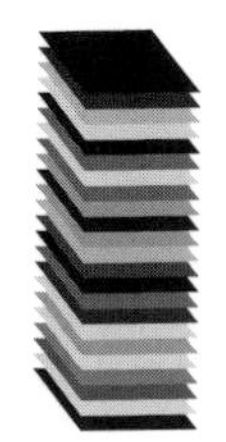

The Bridge Tour

There is nothing quite like the view of a great suspension bridge. Sewage treatment plants are also terrific engineering achievements, but they are not as aesthetically pleasing.

Many years ago, before our son was born, we decided to take a really big vacation. Everyone said that once you have kids, life is different. It turned out that everyone was right. My wife wanted to go to Europe. We settled on a few weeks traveling around England. She wanted to see such lesser attractions as Big Ben and the Queen. I had other sights in mind. I had mapped out a route to see the great bridges of Great Britain.

We drove to Wales across the original Severn Bridge, a beautiful suspension structure with a graceful aerodynamic deck and slanted suspenders. To the north, over the Menai Straits, we visited the Menai Bridge and the Britannia Bridge. The Menai Bridge is one of the world's oldest suspension bridges, predating the Brooklyn Bridge by 30 years. It was designed by the renowned British engineer, Thomas Telford. The Britannia Bridge had unusual, massive hollow-box-shaped plate girder spans that trains would drive through. The Menai Bridge has largely survived in its original form, but the Britannia Bridge has been built and rebuilt due to fire and reconstruction.

Next, it was on to Scotland and Edinburgh, a great medieval city featuring one of the world's great civil engineering sights. You stand on the shores of the River Forth between two spectacular structures: the Forth Road Bridge, a long modern suspension span, and the Forth Rail Bridge, a hulking, massive towering double cantilever. The railroad bridge was built after the Firth of Tay bridge disaster. On December 28, 1879, a gale blew out some of the Firth of Tay bridge truss spans, and a train carrying 75 people plunged into the river. British and Scottish engineers responded to this tragedy by taking no chances with the later Firth of Forth design. The resulting structure is monumental and somewhat over-designed. By its appearance, clearly the rail bridge will not be tipped over by the wind.

To complete our tour, we visited the city of Hull. Apparently, this is off the beaten track and not high on the tourist list of places to visit. But for civil engineers, Hull offers the Humber Bridge, a huge suspension span with tall concrete towers and angled suspenders. When I saw it, the Humber Bridge featured the world's longest span. Of course, records like this are made to be broken.

Although my wife might not necessarily agree, I think there are so many bridges to see and so little time. Fortunately, to supplement vacation visits, we can attend lectures. At an ASCE convention in Boston, we had a presentation on the Akashi-Kaikyo Bridge. This Japanese suspension bridge is the new record-holder for the world's longest span. It is a truly massive structure, an incredible leap of a span, more than a mile and a half between towers. The engineering for this had more than the usual share of adversity. The structure is near the city of Kobe, which was severely damaged by an earthquake in 1995. At the time, the bridge towers were up but the span was not constructed. The towers shifted several feet relative to one another as a result of the earthquake, and the structure had to be reevaluated. The Japanese authority maintains a Web site (http://www.jb-honshi.co.jp/english/index.html), which includes descriptions of this bridge and other structures that cross the inland sea.

In the future, new spans will be longer than the current record-holder. Some models have been prepared for proposed spans across the Straits of Gibraltar and for connecting Sicily to mainland Italy. Unfortunately, the models feature ungainly spider-web assemblages of cables and not the graceful arc of the traditional suspension bridge. For shorter spans, cable-stayed bridges are increasingly proving to be more economical than traditional suspension bridges. Cable-stayed bridges can be sleek and elegant, such as the beautiful Tatara Bridge in Japan and the new Zakim Bridge in Boston. But in my mind, cable-stayed bridges are not as graceful and grand as suspension bridges.

The traditional suspension bridge span is an archetype and a symbol of the modern city, at least of the modern city of the past. The image that comes to mind of the bustling city is that of Manhattan, with its many suspension bridges. This image has been played out in many graphics and films. Manhattan is often visually represented by the backdrop of sky-scraping towers with the Gothic Brooklyn Bridge crossing the East River in the foreground. Many of us are familiar with this view, although now it is tinged with sadness and anger at what is no longer in the backdrop.

Sleek suspension bridges are not just icons of urbanity, but also representative of bridge disasters. Many early suspension bridges failed. Before the successful Brooklyn Bridge, John Roebling's Niagara Bridge near the falls was blown down. Yet, no bridge has failed so spectacularly or photogenically as the infamous Galloping Gertie, the nickname for the first Tacoma Narrows Bridge in Washington State. This bridge was built with a slender plate girder and a narrow two-lane deck. During a gale on November 7, 1940, the deck started to oscillate about its center line. Many of us have seen the film clip of Galloping Gertie's last ride, frozen in time in a newsreel loop. The span twists and writhes, and then rips itself apart and plunges into the Narrows. Except for a reporter's dog trapped in a car, fortunately no one perished in this disaster, but the collapse affected suspension bridge design in the decades following. You can clearly see the change in the second Tacoma Narrows Bridge. As with the hulking Forth Rail Bridge, engineers took no chances with this structure. The second Tacoma bridge has a deep, boxy stiffening truss. Another, very different design approach is exemplified by the Severn Bridge, with its aerodynamic shaped deck section that attempts to reduce dynamic wind loading instead of stiffening against it.

Today's cable-stayed bridges are being built to increasingly longer spans, so maybe the days of the traditional suspension bridge are numbered. But that trend could be bucked by another trend bringing back suspension bridges. Two of the most recent large U.S. bridge projects are for suspension bridges. Both are replacing old cantilever truss bridges across San Francisco Bay. The first is the new Carquinez Strait bridge, a suspension span to take the place of the older of the two cantilever bridges currently there. The second project is the unique mono-cable for the east span of the Oakland Bay Bridge, a replacement for the cantilever truss span that failed during the 1989 Loma Prieta earthquake.

In 2002, the new Natcher Bridge was completed. The structure spans the Ohio River at Owensboro, Kentucky. The bridge is very beautiful and dramatic. The structure looks sleek from the river, a near-perfect expression of what I think of as "bridginess," the ability of a bridge to express its structural form and relation to what it spans. The pylons are grand and formal, in the tradition of great suspension bridges such as the Bronx Whitestone and Verrazano-Narrows. The new Natcher Bridge takes a place on the list of the world's most beautiful bridges. It helps to make that stretch of the Ohio River someplace to visit.

In the 19th century, upper-class young men took a Grand Tour to put the finishing touches on their education, rounding out their world view by visiting the great monuments of Western civilization. Maybe today, young civil engineers should make a point of doing something like a Bridge Tour.

Fred Retires

It was a much better day than the average Monday. On a warm, September day, the doughnut company decided that its advertising icon would retire. After 15 years of hawking doughnuts in Boston, bedraggled Fred would hang up his skillet. To commemorate this event, the company offered free doughnuts and coffee at its shops. This was quite a sight to see: the long lines spilling out onto the streets, the Bavarian creams flying from the shelves, the customers bouncing off the walls from extra caffeine and sugar highs. In general, I think that free food is a good thing. On this day, I was fortunate that my job took me to many outside offices, past several doughnut shops. I got more than my share of the celebration. If only they had had the foresight to offer free bagels as well, almost all of the major food groups would have been covered.

As I was munching on a chocolate doughnut, I started worrying about Fred's retirement. What would Fred do now? For 15 years, he has dragged himself out of bed at what looked like 2 a.m. Doughnut making must be complicated if you have to wake up that early. Although Fred appeared devoted to his work, I don't think he really enjoyed it. He looked a bit oppressed all those years, whining about the time it takes to make the doughnuts. He seemed genuinely relaxed to have the burden of doughnut making lifted from his corpulent frame. Maybe Fred would putter around the garden or play some golf. Probably he would have an omelet for breakfast, but not doughnuts.

When a prominent civil engineer retires, it's not as big a deal as with Fred. You don't see the city coming to a halt with hordes of people waiting in line to receive free calculation pads and pencil lead. Engineers lead quieter lives than those in the more prestigious profession of doughnut making. Engineers tend to be more introspective and not as much in the public eye.

There is a big difference between engineering and making doughnuts. Frying doughnuts is a mechanical activity. Engineering, however, is not just an activity but a way of thinking. I bet it was easy for Fred to hang up his apron each day and leave the job at the office. He probably didn't think about doughnuts while, say, showering or taking out the trash. Engineers, however, don't stop thinking about engineering when they leave the office. We want to fix the shower head, improve the water flow, and optimize the space in the trash barrel. We engineers have trouble escaping from being engineers.

A funny Ann Landers column used to hang on a colleague's office door. The column concerned the opinions of beleaguered spouses married to engineers. The letters were about half positive and half negative. Some of the spouses appreciated the logic and reliability of their engineer husbands or wives, in comparison to, say, a used-car salesperson. Things always got fixed around the house. Other spouses decried the cold logic and lack of passion in their mates. Passion is not easily quantified or optimized, so it can be tough for engineers to grasp the concept.

If engineering is not just a job, and not just an adventure, but this all-encompassing, overarching experience, what is an engineer to do when it's time to officially stop?

The Sky Bridges and Malls of Minneapolis

At the October 1997 ASCE convention, the civil engineers all converged on downtown Minneapolis. It was my first visit to the city. I didn't know what to expect. My knowledge of Minneapolis included Mary Tyler Moore throwing her hat up in the air (making it after all), extreme cold, and something about a really big shopping mall. During my visit, it was uncharacteristically warm and sunny, so I didn't get a taste of the infamous Minnesota deep freeze. However, the severe winter climate certainly had an impact on the city I was about to see.

It turns out that Minneapolis is a much different city than Boston. Two features made a big impression on me. One was the really big shopping mall. It's called the Mall of America, and it is really, really big. It is a local tourist attraction. I met a West Coast colleague, and we took a bus out to the mall for a few hours for inspection.

The mall is situated to the south of downtown, near the airport. You arrive at a miniature bus transit station, and climb up the escalator to enter the mall. The building has three levels spread out over many directions. In the basement is an aquarium (a big aquarium) with octopuses and a walk-through shark tank. The mall has several atriums. The really big central atrium contains an amusement park, with a water flume ride and a roller coaster that winds its way around the shops. Befitting the scale of the mall, there is a really big food court, with tables and terraces that look out over the roller coaster and the amusement park. I doubt we saw all of the mall, because it's big, but I noticed things like a giant Lego exhibit and a tour of cars of the future.

The Mall of America is an impressive place. Ultimately, however, it is just a shopping mall, even if it's large. As a shopping mall, it features all of the troublesome urban impacts posed by shopping malls. This mall is surrounded by the typical suburban wasteland—a big parking lot. The mall is not part of the fabric of any city but is separated from it as a walled-off fortress. Inside the fort is a very pleasant but static and controlled environment. Unlike a real city, this imitation city only has shopping (with fish and a roller coaster). At certain times each day, all life stops and the building is emptied. As in most shopping malls, the experience is controlled and numbing. Being at the Mall of America is not like being at Rowes Wharf or Post Office Square Park or almost anywhere in the more vital streetscape of downtown Boston.

The second feature of Minneapolis that made a big impression on me is the skyway. This is a system of pedestrian bridges downtown that link most of the major buildings. Unlike the pleasant weather during my visit, Minneapolis is not that great a place in January. They show you Mary Tyler Moore throwing her hat up in the air. What they didn't show you is the hat freezing in mid air and not coming down due to the extreme cold. Urban planners have addressed this bit of adversity by designing the system of pedestrian bridges. The skyway winds between buildings at the second story level. In the winter, by using the skyway, it is possible to drive to downtown, park, and pretty much never step outside. You go from place to place in heated comfort as the Canadian Arctic wind rips down the plains outside.

Unfortunately, the skyway has had an unintended and unpleasant effect. It has led to a sort of mallification of the entire city. I arrived late on a Saturday night and wanted to get a quick bite—maybe a bagel or a muffin. I walked outside my hotel onto the street and saw nothing: no people, no shops, no food, just a bunch of mostly blank building fronts and garage entrances. There was almost no life at the street level. All the action had been sucked by the skyways to the second floor of the buildings. Here you could find food stores, dry cleaning, clothes shops, but it had been turned from a normal city street into a linear shopping mall, with restricted hours and the usual sterile mall conditions. The street level was largely a pedestrian desert intended only for cars and parking. The city had introduced one feeble attempt to generate some activity on the street. One road in this perfectly square grid city had been given some curves, and a few shops and stores bravely put out some chairs and tables on the sidewalk. They looked out of place.

Visiting Minneapolis and other cities helps you to appreciate how fortunate we are in Boston. The city has winding, quirky streets, buildings of different shapes, sizes, and ages. The streets are active and pleasant to walk on. There are all sorts of different activities, reasonably civil but not controlled like in a shopping mall. Boston does have its own, smaller version of the skyway, a sort of mini-Minneapolis. This skyway is a group of pedestrian bridges connecting Copley Place to the Prudential Center and Westin Hotel. The effect of the bridges is clear to see. Huntington Avenue below is a speedway, and the cityscape at street level is barren there. Compare this area to the Back Bay just a block or two away.

The lesson from this seems to be that separating the pedestrians and activities from the street is not a great idea. Maybe Boston winters are not as bad as in Minneapolis, but they are still plenty cold and uncomfortable. Fortunately, Boston has managed to preserve the streetscape and fabric of the city, and even enhance it. The cliché is that Boston is a "walkable" city, and what this means is that it is lively, architecturally diverse, with a working downtown that hasn't been mallified. With this in mind, ideas for "traffic calming" and moving traffic underground are probably better than walled-off shopping malls and skyways that separate the people from the city.

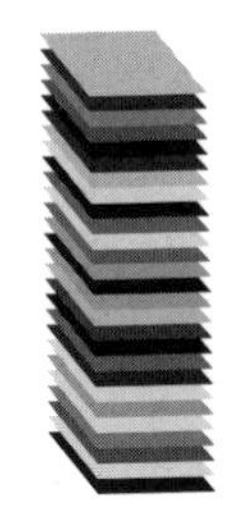

Raising the Bar

There is a guy on my block who is the most technologically advanced. He supplies the peer pressure for the rest of us. He has nine PCs in the house. Since only four people live there, not counting the two gerbils, that's 2.25 PCs per person. The PCs have been networked. Well before the days of easy wireless, my friend laid and strung some sort of cable (I think it has the letter "T" in it). The cable winds through walls and ceilings and at one point pops out of the outer wall to climb upstairs. Some of the other PCs are wireless, which is to say that they communicate via a type of microwave connection. If you stand between the wireless talking PCs and put a film plate next to your mouth, you can also check for cavities.

I wondered if it was really necessary to network the PCs. My friend would proudly exclaim how he had attached all of the new technology and gizmos. For example, his kids no longer had to wake up to print a spreadsheet. They could imagine it, and the neuroimplants would send it through the wireless. This is technologically possible, I think, so it's only a matter of time. For me, a home PC network is a hypothetical possibility at best. We have only one PC. We do, however, have many toilets. To compete with my technological friend, I imagined that I would network the toilets. What possibilities that would offer! You could flush the upstairs toilet from downstairs.

I suspect that there is a point when enough is enough. A reasonable cost–benefit analysis would show that the effort you spend putting up with the new technology greatly outweighs what you're getting back. To be truthful, my friend is a bit of a nerd. He enjoys the challenge of hooking up the wires and playing with the software. He likes to get a computer virus because then he can reload all of the programs. But for the rest of us, the need to be out on the cutting edge competes with getting cut by the cutting edge.

The wave of innovation seems to roll in fits and starts. One brave bunch is always ahead of the curve, venturing out into the unknown frontier. Sooner or later, the rest of society catches up and the frontier becomes settled. The restless then move on to conquer some new technology, some new unsettled Wild West. But the technological edge has many dead-end alleyways, and for each good innovation, dozens of methods and approaches didn't need to be invented in the first place. Those of you who step a few feet back from the edge perhaps aren't so brave and adventurous, but you live longer.

The AASHTO Load-Resistance Factor Design bridge specification illustrates my point. This relatively new bridge code is at the technological cutting edge of American bridge analysis and design. Consider the calculation of the live load distribution factor for bridge stringers. This is the factor that allows you to distribute a part of a truck load applied from the bridge deck down to the individual stringer. In the old code (English units), for steel stringers spaced at distance, S (in feet), the live load distribution factor was defined as: $S/5.5$.

Figure 1 illustrates how you're supposed to do the calculation in the new LRFD code. The new equation has six variables: stringer spacing, plus deck stiffness, plus girder stiffness, plus span length. At least there are no differential equations, but plenty of quadratics and exponentials. When you're done with all that, you probably end up with a range from $S/5.4$ to $S/5.6$. It's not clear what this more precise calculation achieves in terms of bridge design, other than taking up more of your time and calculation ability. Perhaps in a few years, the LRFD code will become more transparent and less involved. When that happens, the new code will become the old code. I suspect that then someone else will be working on stringer distribution factors that use imaginary numbers, nanoparticles, and quantum theory.

The cutting edge of technology is messy and features other dangers for you to be wary of. To best apply technology, you need to step back from it and think about what you're doing. In the example above, the old, boring method is reasonably understandable to apply and straightforward: the further apart the beams are, the less load they get. Using the new method, which is closer to the cutting edge of distribution factor technology, things aren't so transparent. The true meaning of the new equation is hidden by its complexity and its relative difficulty of use. You spend more time figuring out how to apply the equation correctly, and less time thinking about the bigger picture of what it all means. So, at the technological edge, there's no stepping back, because you have to concentrate on not falling off.

I have great appreciation and admiration for my technological friend. He and his colleagues out on the frontier drive the technology forward and eventually make life better for the rest of us. My friend figured out how to network his house years ahead of the average American. His kids are probably better off for it, at least when their computer system doesn't crash. In my case, I am content to drift slightly behind with my one PC, only recently upgraded from DOS. When the peer pressure becomes too intense, I'll save face by showing my friend how to network his toilets.

For two or more design lanes loaded, in SI units:

$$0.075 + \left\{ \left(\frac{S}{2,900} \right)^{0.6} \left(\frac{S}{L} \right)^{0.2} \left[\left(\frac{K_g}{L} \right) t_s \right]^{0.1} \right\}$$

where

S = stringer spacing (in mm)
L = span length (in mm)
t_s = thickness of concrete deck (in mm)
K_g = the longitudinal stiffness parameter,

$$K_g = n \left[I + (A \times e_g^2) \right]$$

where

A = cross sectional area of stringer (in mm^2)
n = ratio of modulus of elasticity of steel to modulus of elasticity of concrete
I = moment of inertia of stringer (in mm^4)
e_g = distance between centroid of concrete deck and centroid of steel stringer (in mm)

Figure 1. LRFD Method of Calculating the Live Load Distribution Factor for Bridge Stringers

Vegetarian Nerds Watching the Super Bowl

In 2005, when the Patriots didn't make it to the Super Bowl, I lost interest. We had been spoiled in Boston with a remarkable streak of Super Bowl wins for three out of four years. But when the streak came to an end, for many in New England, the Super Bowl seemed pointless.

Even though the game was pointless, my technological friend wanted someone to watch it with on his big TV. His TV is a device with a high-definition screen, surround sound, and extensive attachments. To turn it on, you don't just flip a switch. You type a list of commands on the complex keyboard. My technological friend is very proud of the keyboard because it can do many things. After about 10 minutes of fidgeting with the commands to accomplish things (I'm not sure exactly what), he finally gets to the command of interest: turn on TV. The sequence is so complicated that it takes another five minutes for the device to warm up. In comparison, my ancient TV at home is a little easier, having one "on" button and no activation keyboard. It turns on within about three seconds. Then you have to adjust the bunny ear antennae for proper reception because I don't have cable.

So I didn't really want to watch the game, but I didn't want to let down my friend either. Besides, the TV (once it's finally on) is really impressive. The speakers are so well placed behind where you sit that it seems like the sounds from behind you are coming from behind you. The layout of the TV space is perfect, featuring a slouchy array of comfy couches and footrests. There is probably no better place in my neighborhood to goof off than this TV room, and the Super Bowl was as good an excuse as any. Also, we are both Type A++ personalities, his being extreme technological, and mine being Luddite, so any excuse to goof off while slouching in the comfy sofas is something to cherish.

Other than admiring the TV, however, I should admit that watching football is not an activity that we excel at. This is because we're nerds. For starters, my technological friend is a vegetarian. Normally when you watch the Super Bowl, you're supposed to have roast ribs and other kinds of flesh. All over town, and probably the rest of America too, viewers made mewling sounds while gnawing on remains of dead animals. In this primeval scene, as the pigskin chasers pounced on the flat screen TVs, streams of barbecue sauce dribbled down unwashed, stubbled chins, mixed with beer.

We had some barbecue sauce on our chins, too, but the roast ribs were made of tofu. It's really not the same.

Football games have a lot of new technological gizmos to interest nerds, even vegetarians. There's that computer-generated yellow line on the field that tells you where the first down is. This is neat stuff—the computer figures out graphic interferences with the players and moving objects in real time, so it looks like the yellow line is painted on the field. I would say to my son, "Dan, look how they painted the first line marker on the field," and he would roll his eyes. Today's first-down demarcation line on TV is presented in sharp technological contrast to the old way of doing things—the chains. On TV, you see a solid yellow line that conveniently moves with the football. On the actual playing field, there's no yellow line, and it's back to the 1950s as the mechanical first-down chains are physically moved back and forth. With all of the nanotechnology available, you wonder how the referees can get it right, or even measure it at all, when they determine that there's an inch to go for first down. Maybe the measurements are somewhat subjective. Maybe the technology doesn't need to be that precise.

Even with all the opportunities of instant replay and precise precision, things often get a bit subjective. When we weren't watching the Super Bowl, we saw the finals of the national spelling bee on ESPN (and I'm not sure how a spelling bee qualified to be on ESPN, but that is a digression for another time). In this competition, middle school students would be challenged to spell difficult words, and the five judges would decide if their renditions were correct. A simple exercise to judge, you would think, not much ambiguity here. For one particularly difficult word, a seventh-grade boy clearly stated the letter "e" when it should have been "o". He was toast. But the judges took about 10 minutes to determine that the word was, in fact, misspelled. They carefully conferred and debated while the TV showed and reshowed instant replays of the action. We watched the poor boy sweat and resweat, all from different camera angles.

Clearly, camera shots at different angles are of value. When you watch the Super Bowl, the TV shots are great, probably a lot better than anything you'd see live in the stadium. The best camera is held up by a complicated tension structure that suspends the device directly above the field. The camera is held by a series of transverse and longitudinal wires that provide support, but also move the camera along the playing field. The result is a series of sweeping, swooping video shots that move forward in the direction of play. The halfback runs with the ball, and you seem to be running with him as the suspended camera lurches forward. This is pretty impressive, especially the design of the wires.

It is interesting to think of football as an exercise in engineering. When I attended MIT, they started a football team. MIT wasn't known for excellent football then, and I think this is still true today. When it came time for football cheers, the students didn't know what to do. One student in the stands called out:

"Give me an M."

We spectators responded with an "M."

He cried: "Give me an "A."

We responded with an "A."

He cried: "Give me an "S."

We responded with an "S."

Twenty-five minutes later, we spelled "Massachusetts Institute of Technology." By then the players had left the field, and it was part way through halftime.

Sadly, not a lot of engineering input and mathematics were involved in cheering, so the students decided to get creative. Before a Harvard–Yale football game, MIT students buried a black, uninflated weather balloon in the field. During the second quarter, one student flipped a switch. The balloon started inflating, and it popped out of the ground, stopping play. Classic photos in the MIT Museum show the delighted, bemused, and horrified looks of the football players and fans as the giant black balloon, decorated with the letters "MIT," inflated and eventually burst.

Back at my technological friend's house, the Super Bowl was almost over. One of the two teams was in the lead. My friend offered me some slabs of leftover tofu, which were congealing in the cold barbecue sauce. A funny commercial came on, something about a man working in an office full of chimpanzees. All over America, spectators were winding down their Super Bowl viewing ritual. Many would quickly turn their TVs off. But in our case, it would take a complicated command sequence on the keyboard and another 10 minutes before the lights on the TV dimmed.

My New Cell Phone

For many years, I resisted getting a cell phone. I was afraid that it would be too distracting if it rang every other minute. Also, I thought I would lose it within a day or two. Cell phones are little devices that easily fall out of my pocket. The cell phone company said that I could get one to attach to my belt. But this seemed like a modern update to the days of wearing slide rules. First, I would wear a cell phone on my belt, and then I would need to get a pocket protector and discontinue personal grooming habits.

With my new position at Tufts University, the more complicated commute and schedule pretty much required that I have a cell phone. So I bit the bullet and tried to adjust. When I first got it, I put the phone in my pocket and drove to a neighbor's house that evening. Sure enough, the thing started ringing and vibrating the minute I was on the road. This was my big fear—to be distracted by the cell phone while trying to do something else. I tried to answer the call. I kept one hand on the steering wheel and with the other maneuvered my cell phone into an operable position. But sadly, I learned that it is not so easy for a beginner to answer the phone, since there were many choices and buttons. I'm not the greatest multitasker, so I focused on driving to avoid a multicar accident.

When I got to the neighbor's house, I saw that I had a message, but it was difficult to figure out what it was. Apparently cell phones are more complicated than traditional phones, which in the good old days used to have two functions: listen and talk. My cell phone had a menu with dozens of options. I could answer calls, send e-mail, play little cell phone games, and even cook dinner (something called an "infrared" option, which I assume is used to cook dinner). I explained to my neighbor that it was my first cell phone and that this was the first time I was using it. The neighbor thought I was joking. After about five minutes, she incredulously saw that I was serious and that it really was my first cell phone. She took pity on me and walked me through complex menus. Eventually we got to level 17, where the messages were. The message I received, the message that caused the buzzing and ringing while I was driving, was from the cell phone company. They were calling to congratulate me on my new cell phone and to welcome me to the wireless world.

In retrospect, the company's friendly call wasn't so bright. If a lot of the new customers had my skill level, there was a big potential for accidents on the first day of use. Those friendly welcome calls could end up killing off a new revenue stream, literally. The subject wasn't such a joke, either. We now have to face the problem of cell phone auto accidents. Even experienced users have trouble manipulating the phones while driving. Some states have written laws prohibiting use of a handheld cell phone by people who are driving. The topic is the subject of serious transportation research—how to talk and drive without getting killed.

The little cell phone introduced a new element into my daily routine, a routine that unfortunately is somewhat rigid and not accommodating of new elements. Now I have to remember to remove the cell phone from its charger and stick it in my pocket. To date, I have usually forgotten to take the cell phone in the morning. Apparently I've reached my limit for remembering morning activities. It's the theorem of Morning Activities Equilibrium. If I remember to take the cell phone, then to satisfy equilibrium, I must forget some other important thing, like putting on underpants or combing my hair. A few weeks ago, for example, it was important that I remember to take the phone during a trip to New York City, and of course I forgot it (but I did put on underpants). Then while in New York, my mother called to change plans using the new cell phone number (what that is, I have no idea). Unfortunately, she couldn't get in touch with me, because the cell phone rang upstairs in the bathroom where I had left it.

Like other modern gizmos, cell phones are making their presence felt on the staid world of infrastructure design. Cell phones need special towers to enable the communication. To ensure uniform coverage, the towers need to be placed everywhere, including areas not used to having large towers. These tall structures have resulted in a new wave of not-in-my-backyard issues. For example, in my hometown, there was recently a big fight when the local golf course agreed to allow the cell phone company to build a tower. The neighbors protested that it would be big and ugly. They tried to refer to the rules, but the whole issue is so new that there weren't any. Having a zoning fight is difficult enough with rules, but without them it's the Wild Wild West.

To avoid problems, some cell phone companies have gotten creative with their towers, camouflaging them to look like trees or hiding them in church steeples. In Tucson, one tower is hidden in a sculpture of a really big cactus. This is perhaps ironic, because some studies have shown that the giant Saguaro cacti are gradually dying off, maybe due to the cell phone radiation. If you look carefully, you can see the peculiar-looking cell phone tower-trees along freeway rights-of-way. But you shouldn't try to drive, look at the funny trees, and answer your cell phone at the same time.

My cell phone has that boring beginner's ring that all new users get. I suspect that eventually I'll master the technology and start to explore the nuances. Then I'll use it to store addresses, take pictures, and even cook dinner with the infrared option. At that time, I'll program the phone so that it will no longer ring with the boring beginner's song. It will play *Also Sprach Zarathustra.* For the privilege of speaking to me, callers will need to listen to 15 minutes of this before I answer, and a hologram of an ambiguous black monolith will orbit their heads during the wait. And sure as you can spell "DVD," the moment I learn to accept my cell phone and even embrace its complexity, the moment I become comfortable with it, some new disturbing technology will come along to shake my Luddite ways and save my life. Then I'll have to learn nano-faxing, or personal barbecuing, or whatever the next great distraction turns out to be. Because whatever it is, it will be something we can't do without.

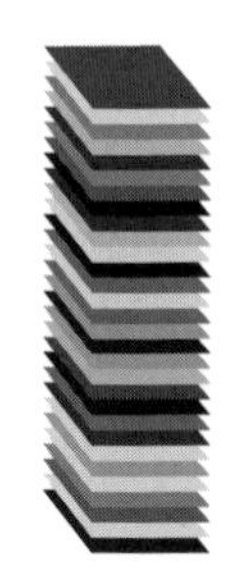

Hamsters Gone Wild

We're not the best family for pets, having killed many goldfish and two cockatiels. One time I had a parakeet named Clyde. Before you hear the rest of the story (so to speak), I should point out in my defense that this was a really mean parakeet. We were not the first owners. The bird was originally given as a gift from a man to his fiancée. In this case maybe flowers would have been better, because the man left her at the altar. She expressed her anger by torturing the parakeet. It was a reverse "he loves me, he loves me not," done with feathers instead of flower petals. Eventually she couldn't stand doing even that, so she sought to discard the bird. We wanted a pet, and this was before we understood our deficient pet-rearing abilities, so we rescued him. But by then, Clyde was pretty much finished dealing with human beings. As a result, Clyde was not easy to love. It wasn't like the cherubic birds in *Mary Poppins* that would chirp along with Julie Andrews and land on her finger. No, Clyde was out for vengeance.

One time I was cleaning the cage while Clyde was still in it. I carefully avoided his nasty beak and sharp talons. I used the vacuum cleaner with the top attachments off. I got distracted, and then I heard a "thunk." Poor Clyde was stuck at the top of the vacuum, his little feet and tail feathers wiggling out the end of the hose. For about a week or so after that, Clyde was a bit friendlier, or at least not as nasty. It was either because the cage was cleaner or the bird feared being vacuumed again.

We haven't had cats or dogs, and we won't in the near future. But my daughter loves animals, and she tried for years to get another pet in spite of our dysfunctional abilities. Finally we relented and got hamsters. This seemed like a safe choice. Hamsters stay in cages, and they're really mice, which should have been pretty hardy, even in our household. Also, I was frequently reminded by my family about the time I vacuumed the parakeet, so a hardy pet would be preferable to a more delicate creature. We thought about getting one hamster, but that seemed wrong because a solitary mouse would be too lonely. We got two, with the pet store assuring us that both were girls. My daughter named them Wolf and Zeppelin.

In the beginning, both hamsters were adorable little babies that needed help drinking water from the hanging dispenser. However, the life cycle of a hamster is much shorter than that of a human being, or even a dog. Within a few weeks, Wolf and Zeppelin were much bigger. In fact, Zeppelin grew a surprising amount in a short time. We thought it was from too much food and watching TV with my daughter, but there was another cause of the sudden weight gain.

Instead of two girl hamsters, we had one girl and one boy. On a Friday morning, my daughter shrieked—"Dad, you have to come see this!" A gaggle of tiny babies squirmed in the cage. We had a hamster family of 11. How the two original hamsters had gone so quickly from being babies to parents themselves, I'm still not sure. I thought of an episode of the original *Star Trek* series, in which the crew encountered cute, fur ball creatures called "tribbles." The trouble with tribbles is that they reproduce, continuously and often. Dr. McCoy concluded that tribbles were essentially born pregnant, close to the condition of our pets, who had a few weeks of hamster childhood. Dr. McCoy's key advice was that you shouldn't feed the tribbles. This didn't seem like an option for our pets.

With the now expanded family of rodents, we had to deal with all sorts of hamster infrastructure issues. The cage we bought was pleasant and spacious for the two pets. The split-level cage had a health club spinning wheel on the side and a plastic house on the top level where the two hamsters could hide out and rest. This was where Wolf and Zeppelin decided to raise the kids. What was cozy for two little hamsters seemed pretty crowded for two big fat hamsters and nine growing babies. Apparently hamsters like to smoosh all together.

Watching this growing family, I thought about issues of sustainability. In the microenvironment of the hamster cage, things would quickly get out of hand. In a few weeks, we went from two pets to 11. Assuming the same rate of growth, with some conservative assumptions, we would have about 900,000 hamsters a year later. Clearly we would be out of room well before then. To support even a month or two at the current growth rate, we would need a vast expansion of the hamster infrastructure. We would need more cages, more hamster houses, more food, and more water bottles. We would need lots and lots more hamster exercise wheels, and they would all be spinning furiously at 3 a.m. because the creatures are nocturnal.

Something would happen before we reached the one-million-mouse mark. The initial growth rate would not be maintained. Another thing we learned about hamsters is that they have problems with maternal instinct. After a few weeks of loving care, the mother hamster started eating the babies. One by one, the kids turned up dead in the cage, until my wife actually caught Zeppelin in the act. At that point, Zeppelin was banished to a hamster ball and eventually back to the pet store. This type of behavior is not unusual in hamsters, but of course we were clueless.

The two extremes of hamster behavior help frame the debate about sustainable development for human beings. One extreme argument postulates that our human future involves massive overpopulation and environmental degradation, with apocalyptic results. For example, a global warming disaster movie in the summer of 2004 showed giant tidal waves and tornadoes blowing away Hollywood. At the other end of the scale, a more laissez-faire approach is endorsed. The argument maintains that things will readjust by themselves regardless of what we do. It's not that human mothers will start behaving like hamsters to reduce the rate of growth, but factors that we are not taking into account will moderate the doomsday scenarios. One theory holds that the vast reservoir of energy and mass in the ocean naturally moderates any changes we might cause.

There is evidence to support both scenarios. Rapid human growth and infrastructure development have probably led to some global warming, with uncertain impacts that we are unable to model with great accuracy. On the other hand, birth rates have decreased drastically in the last decade or so. The vision of *Soylent Green,* a cheesy post–Moses movie in which Charlton Heston lived in a failed, overpopulated world, now seems like an overwrought and incorrect extrapolation of the future. The debate is often phrased like the classic engineering approach of alternatives analysis. Engineers as problem-solvers must do something, even though doing nothing is one of the alternatives. Maybe doing something could be a worse alternative. On these lines, for example, assigning a limit to the human population has the negative impact of fewer people in the community, which can be a bad thing: what if one of these unborn individuals would have discovered the cure for cancer or the follow-up to string theory?

Civil engineers are now grappling with the difficult set of questions related to sustainability. What is needed is a cross-discipline, big-picture understanding of problems and interrelationships. For many engineers used to a more atomized, discipline-specific evaluation, this requires a new set of skills and a new approach. Another challenge involves comparing and weighting entirely different issues—for example, the economic bottom line of a project versus projected social impacts 20 years from now. Our current approach is not much better than formulating checklists of what's been accounted for and what hasn't. The old bottom line still tends to carry the most weight. But true sustainability is what is demanded of our projects now and in the future. Our evaluation, analysis, and design tools must rationally address this requirement. Engineers need to get ahead of the learning curve, or we will be continuously in reactive mode.

As with most complex problems, the best responses are often found closer to the middle and not at the extremes. As it goes with infrastructure design, so it goes with pets.

A Comparison of Dilbert and Wally

The goal of this essay is to present a comparison between Dilbert and Wally, two characters in the comic strip *Dilbert* by Scott Adams. This important evaluation has been needed for a long time. Since all of the characters are engineers, except for the Boss, who is Management, we engineers can learn important lessons from the comparison. Perhaps such a comparison will lead to improvements in our own engineering contributions and productivity, a goal the Boss would appreciate.

But first, a digression:

There is a scene in the movie *Animal House,* in which the college freshman hero and his sidekick are going to visit fraternity row. They go to a party at a serious, snooty fraternity ("Delta Delta Something"). The host at the party takes one look at the two guests, nicknamed the Wimp and the Blimp by a sniggering fraternity member. Then, the duo is shown to a corner where the host thinks they will be more comfortable. In the corner is a group of what appear to be social misfits, guys with questionable hygiene habits, dubious taste in clothing, and perhaps limited conversational skills.

Sometimes I've thought that this scene lacked only one thing to be complete and authentic: a large banner reading "Engineers" hanging over the group.

OK, now on to the main topic, a comparison of Dilbert and Wally.

But first, another digression:

For a while, I wondered why Dilbert's tie curved up. You will notice that every tie Dilbert wears is curved up like a fish hook. Until recently, I didn't understand why. Then, one day, I looked at some old ties in my closet. It turns out that if you keep a tie long enough, the fabric stiffens and it curves up. I had empirical evidence. My ties were still wearable, of course—no obvious stains, structure of the fabric still intact, more or less. It is the engineer's job to maximize materials and optimize performance. These were still perfectly viable ties that had a few seasons to go before their visit to the Salvation Army. But they curved up, like Dilbert's. It was no joke.

OK, now for the comparison.

Dilbert and Wally are two engineers working for an unnamed Boss at an unnamed engineering company in some city, some place. Their circumstances and personalities appear to be similar, but there are subtle, important differences:

- Dilbert has a social life, sort of. Dilbert has a girlfriend. He has an abusive pet. He has hair.
- Wally has no social life and is bald. Wally once sent away to the mythical kingdom of Elbonia for a mail-order bride. Residents of Elbonia, as we have learned, wear turbans and long beards and live in waist-deep mud. Wally's bride turned out to be a pig, literally. Their relationship didn't work out. We know this, because in the last scene, Dilbert wants to know what happened to the bride, and Wally offers him a BLT.
- Dilbert naively assumes that there is some order and reason in the world, however slight. He has an underlying faith in truth and altruism.
- Wally has no such pretensions. He is so cynical as to be beyond cynicism.
- Dilbert has no mouth.
- Wally has a mouth, but it's usually puckered, as if he had just eaten a pickle.

So, what's the lesson here? What can we learn about engineering and life from these two characters? I think Mr. Adams presents Dilbert as the hero, Everyengineer, representing all engineers and their efforts to do good work despite a system that often doesn't make a lot of sense. Consider, for example, the time Dilbert had to prepare a project schedule for the Boss. His schedule had one week for the "work" phase, while the rest of the schedule was filled with months of bureaucratic maneuvers. (In the last frame, the work plan shows Dilbert leaping to his death, an embittered shell of a man. If he's lucky, he would land on the Boss on the way down). Dilbert is an expression of the engineer's sense of order and logic butting heads with the anarchy of the world.

Wally, on the other hand, is Dilbert's alter ego. Wally is a caricature of engineering caricatures. Wally does his engineering without really bothering to recognize the existence of the rest of the world. He is happiest that way. There is probably a bit of Wally in many engineers—the desire to curl up with a really good analysis problem and ignore memoranda and staff meetings. It's nice to imagine that Wally will some day obtain social graces. But Wally knows that all that extra energy devoted to grooming and the Hair Club for Men is basically lost time that could be better spent on the computer.

So there you have it: a comparison of Dilbert and Wally. I haven't included a discussion of Alice, the token woman engineer with pyramid-shaped hair and a superiority complex. That is a topic for a future essay.

The Discovery of Pluto

I had the pleasure of visiting the Lowell Observatory in Arizona. The observatory sits atop a hill in Flagstaff, a small city at an elevation of about 7,000 feet above sea level. When the observatory was first built, Flagstaff was much smaller, with fewer street lights and less nighttime illumination. Today, Flagstaff has grown, and the glare from the city is too bright for any meaningful astronomical observation. The work has been relocated to a distant, darker butte out of town in the desert. But the original observatory with its old telescopes and facilities remains for tourists to visit. The site is beautiful, with a pleasant campus surrounded by a southwestern, high-altitude conifer woods. At this site, astronomers discovered the planet Pluto. The methods they used in the 1930s seem primitive today. A comparison between past and present helps to illustrate the great advances in technology that we now take for granted.

The discovery of Pluto took place in two phases. It began with the suspicion by astronomers that there must be a ninth planet, because the solar orbits of the first eight planets weren't quite right. The application of Kepler's Law predicted certain types of orbits, and the measured orbits of the known eight planets didn't match the predictions. Astronomers predicted that another planet was exerting gravity that influenced the orbits of the inner eight planets, and this led to the search for Planet X, the ninth planet.

In the first phase of exploration, mathematicians performed thousands of calculations to try to determine the mass and location of the missing planet. The volumes of the original calculations are on display at the observatory. Today, these relatively simple trigonometric and algebraic calculations can be done on a spreadsheet or with Mathcad in perhaps a few minutes. In the 1930s, it didn't take a few minutes. It took months of painstaking, manual arithmetic, with careful pencil markings, slide rules, and volume after volume of calculations and checking. The mathematicians did all this without computers or expectation that the calculations could be done any way other than longhand.

Phase 1 seemed painful enough to me. At some point, the astronomers determined that the calculations were done, and it was time for Phase 2. Based on the predicted locations, astronomers set out film plates exposed to telescopic images of the night sky to find the new planet. The exposures of each plate were separated by two or three days. The idea was that because a planet would move differently than a star, its image could be identified by comparing images from different nights and looking for a point of light that didn't match up as would a star. The process of comparing the images was done in something called a blink comparator.

The observatory has set up a display of the original device so we visitors could study the method. First, we saw a quick blink of the film plate on the left and then a quick blink of the film plate on the right. Staring into the device, we watched the picture flash back and forth, blink of light after blink of light. We were supposed to find the one point of light out of hundreds that didn't match on both film plates. Gratefully, we looked at a display that had convenient arrows showing the point of light determined to be Planet X, Pluto. With the arrow placed on the map of stars, we easily perceived that one tiny point of light didn't quite line up the way the other hundreds of points did and therefore must be a planet. The convenient arrows were not, however, available for the poor gentleman (instead of "gentleman", I am thinking of a word here that starts with "b") who had to sit for one year with the blink comparator, flashing hundreds of film plates, back and forth, back and forth, trying to find a needle in the starstack. Today, we could digitize the images and have the computer determine the differences in a matter of moments. A year of frustrating, boring, tedious work would now be completed in seconds.

Seeing the exhibit, I thought about how different our lives are today and how changed are our expectations. It's not just in discovering Pluto, but in every aspect of technology and how we apply it. I was preparing to teach a freshman introduction to civil engineering class. I planned to present a lecture on strength of materials. A colleague loaned me an old textbook published in 1951. He had found this book at some rummage sale, purchased for a quarter. It was a musty book but still readable, with good examples and still-relevant themes. The book presented a series of simplified derivations. For example, there was a problem on axial loading of two bars with different cross-sectional areas and different moduli of elasticity. The layout and solution of each problem were predicated on the assumption that you couldn't use a computer. In 1951, even though computers had been invented, they were top-secret, punch-carding, room-filling machines with less computational power than the PC I'm using right now. It took another decade or so before civil engineers could start to imagine practical computer applications like STRUDL and COGO. So, the text was carefully developed with this in mind. The book even had a helpful chapter in the back to assist you with your slide rule.

All sorts of things used to be done this way without computers, whether it was discovering Pluto or designing monumental suspension bridges. Today, this past work seems like monumental drudgery, almost beyond imagination that someone could sit for a year staring at blinking images. We have been liberated by the incredible computation power at our fingertips. It has become second nature, and now the old work-arounds and methods fade into history, to end up as exhibits about finding Pluto and as collections of dusty old slide rules. Considering what people spent their lives doing then, it is so amazing the way we are able to work today. Yet the work-arounds and necessity of the old methods contained a certain grace and embedded level of technical excellence. The manual work was so hard and tedious that there was little margin of error for misapplication. In comparison, at times I wonder about how we use and appreciate the technology that we have been blessed with today.

Postscript

After much debate, astronomers downgraded the status of Pluto from an official planet to a "dwarf planet." This new class of dwarf planets includes Pluto, what used to be the asteroid Ceres, and something that doesn't even have a name (UB 313) but is unofficially known as Xena. So now there are eight official planets, at least three dwarf planets, and all sorts of other bric-a-brac floating around the solar system. The discussion about Pluto's status was reported to be quite emotional (at least as much as astronomical discussions can be emotional). Pluto has a legacy of being understood for decades as being a planet. For many astronomers, downgrading its status was not easy.

What's New on the Xway

For many years, traffic on the Southeast Expressway in Boston has been a case of stuffing the proverbial 10 pounds of material in a five-pound bag. The Xway's fate was sealed in the 1970s, when it was decided not to extend Interstate 95 north from Route 128. The decision not to build the I-95 extension and the proposed inner-belt expressway helped to avoid massive demolition and dislocation in much of Brookline and Cambridge. Instead of what would have been the I-95 Southwest Expressway, the transportation corridor has been transformed into a terrific linear park along the commuter rail tracks. Unfortunately, the existing Southeast Expressway now handles a double load as all Boston traffic from Route 3 and I-95 is funneled to the Braintree Split at the southern entrance. The resulting traffic jams were legendary even before recent reports that almost every other highway in the Boston area has caught up in congestion. Driving from the south, you need to make it to the Split well before 6 a.m. on a typical weekday, or experience a slow roll or worse to the KeySpan Gas Tank and Columbia Road.

Over the years, the scenery adjacent to the Xway has been spruced up quite a bit, to the point where it's actually one of the more aesthetically pleasing highway rides. Driving north from the Split, you pass beneath the new East Milton deck, which has reconnected the village with a park and pine trees. Then it's on to a pleasant tidal marsh and across the Neponset River to the new Pope John Paul II Park. This site used to contain a landfill and drive-in movie theater but is now a beautiful, grassy space adjacent to the river and estuary. Further north, you see the sailboats and sparkling sunlight reflecting off the much cleaner Boston Harbor, with views of tidal flats and sand beaches near Marina

Bay. Even the KeySpan Gas Tank, seemingly an industrial eyesore, has been transformed to an abstract work of art, with its rainbow painting by artist Corita Kent. Beyond that, up to Columbia Road, the decrepit garbage plant with its massive smokestacks has been demolished. At the northern terminus of the Xway, the decaying Mass Pike interchange has been completely rebuilt by the Central Artery project into a series of soaring, sculptured concrete bridges. As part of this work, an abandoned building at the old Broadway Bridge was demolished, and the crumbling bridge itself has been replaced by a nicely detailed concrete viaduct.

Remarkably, almost every mile along the Xway has been redone. Old eyesores and decaying structures have been demolished and replaced. Visitors coming to the city from the south get a much different and much improved visual perspective of Boston, especially with the ugly garbage plant at Mass Ave demolished. It's a good thing that the sites are so much better to look at, since viewers get a close, long look with speeds of 20 miles per hour or less during "rush" hour. Watch the seagulls soaring over Malibu Beach in South Boston, and you can imagine for a moment (or much longer than a moment) that you're at the Cape.

Over the years, the radio traffic reporters have developed a list of markers and shorthand comments to convey just how bad the traffic is. If, for example, the Xway is described as a parking lot from Furnace Brook, then it's best just to stay home. Probably the most prominent landmark along the route is the Gas Tank. This is an appropriate icon for the traffic reports. The tank is filled with fuel, and the cars sitting in the traffic jam burn it off, in a sort of commuting equilibrium (yes, and I know it's liquefied natural gas in the tank and gasoline in the cars, but literary latitude is being taken here). So it was startling in May 2005 to see new competition for icon status when a windmill popped up along that linear shrine to fossil fuel consumption. The windmill was built by the International Brotherhood of Electrical Workers Local 103. It provides electric power to some adjacent buildings.

The windmill project is ingenious on many levels. For provision of power, it's placed at a good site next to windy Boston Harbor. For educational purposes, the project is a winner. Instead of talking about renewable energy, the IBEW went out and built a windmill. As an addition to the scenery, the windmill is quite beautiful. It is a sleek, stark, kinetic sculpture, with a simple cylindrical shaft and white blades. Funding for the project was provided, in part, by the Massachusetts Renewable Energy Trust, which in turn collects funds from electricity rate-payers. In a sense, the fossil fuels are helping to subsidize development of renewable energy sources.

The Xway windmill was planned and built during debate over another, much larger windmill project proposed for Nantucket Sound. For this project, the developer has proposed to build a swath of windmills in the shallow Horseshoe Shoal waters of the Sound north of Nantucket. According to the developer, the wind farm would provide upwards of three-quarters of the electric power needed for Cape Cod, Nantucket, and Martha's Vineyard. The proposed farm would be a nonpolluting, renewable energy source that would greatly reduce demand for fossil fuels. What's not to love?

Apparently, there's plenty not to love. A citizen's group, the Alliance to Protect Nantucket Sound, was formed to protest the project. The crux of the protesters' argument appears to be that the windmills would convert pristine Nantucket Sound into essentially an industrial wasteland. On an earlier version of the group's Web site (http://www.saveoursound.org/), a black-and-white artist's rendering pulled no punches in its depiction of the despoiled seascape, with sailboats menaced by the sinister-looking windmills, beneath a gray, smoky sky which probably contains clouds but is drawn to suggest industrial waste.

Many challenging issues are raised for debate by a proposal to build a wind farm in the Horseshoe Shoals of the Sound. The issues include who is entitled to build in and profit from what is public land (or in this case, water); how facilities like the wind farm should be regulated; how wildlife and, in particular, birds can be protected; and what arrangements should be in place to ensure that the windmills are properly maintained and even removed if necessary.

But of all the issues to be debated, the one that most energizes opponents is concern for visual pollution and potential industrialization of Nantucket Sound. This concern seems slight and unintentionally ironic. It is a slight argument, because windmills are dramatic and quite beautiful, at least in many viewers' eyes. They strike me as less visually intrusive than the behemoth yachts currently plying the waters of the Sound. The concern is ironic, because the people protesting the wind farm are perhaps some of the highest per-capita consumers of energy in the world. I'm not aware of a study to document this, but during a visit to Nantucket, you will see very large homes, very large vehicles on the cobblestone streets, and the enormous yachts cluttering the harbor. Cape and Island residents use a lot of energy, almost all of it provided by nonrenewable sources. Wind power to satisfy these needs would be a good thing in that context. Deepening the irony, one of New England's greatest environmental disasters occurred in 2003 when a barge dumped oil in Buzzards Bay, despoiling much of the coastline. The barge was en route to the Canal Electric Generating Station in Bourne, which generates electricity for Cape Cod.

For alternatives to the proposed wind farm project, the Alliance to Protect Nantucket Sound states on its Web site:

> There are other far more appropriate ways to achieve the same emissions reductions as the Cape Wind project. In fact, given that we do not have a current need for additional power, we should start with an aggressive energy efficient and conservation program. We should also explore land-based wind options prior to going offshore due to economics, risk, and regulatory process.

This is a curious argument, a combination of not-in-my-backyard/there's-no-need-for-it-anyway and a plea for conservation. Even if it were true that there's no need for additional power, it would still be good to replace current fossil fuel consumption with wind power. Opponents to the windmill farm have a good point in the need for aggressive energy conservation. But for now, the SUVs filling the narrow Cape and Island streets keep getting bigger and bigger. A good start would be to replace these vehicles with compact cars and smaller sedans. Some more modest sailboats and dinghies instead of five-story yachts would also be helpful.

Back along the Xway, the white blades from the new IBEW windmill lazily turn in the sunlight, moving a bit faster than the stalled-out traffic. The WBZ copter hovers above, with a dire prognosis for commuters. This morning, it's another slow roll to Columbia Road. But now there's something new and nice to look at along the way.

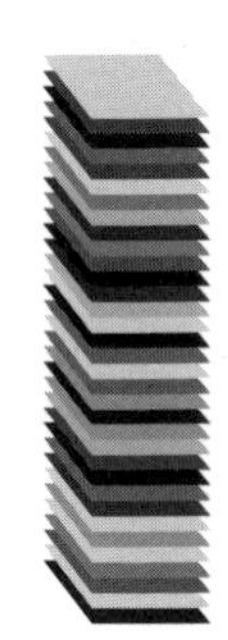

Opryland

The Opryland Hotel in Nashville has some terrific public spaces. It is also emblematic of much of what is wrong with American infrastructure design today.

The Opryland Hotel is astonishingly massive. When you check in, you are given a key map with a list of instructions on how to find your room. The instruction list can have more than eight lines, and it's not just, "Take the elevator at the left to the third floor." The hotel consists of a series of modest-rise wings laid out over several acres. In between the wings, the courtyard space has been roofed over by colossal truss and glass structures. Below these are a series of beautiful enclosed gardens, one more impressive than the next. The first garden has cascading waterfalls, a marsh, a rotating lounge, and beautiful plantings. Another garden has pleasant canyon walkways with fishponds and an Italian patio restaurant festooned with hanging lantern lights. These two are merely the warm-up for the star attraction, a third space that dwarfs the other two. This monumental enclosure has its own lazy river with flatboats, catfish, an island, a geyser water show, and shops with an antebellum theme. Overlooking the river on a bluff is a miniature southern mansion that houses a fancy steak restaurant.

Throughout the hotel are lush plantings—palm trees, junipers—and the sound of running water from the splashing falls and rivers. The architecture is appealing and friendly. Not an inch of the space is taken up by bland. Everything is curving and nookish. The space is designed to comfort, to evoke thoughts of other places, and to just plain entertain in its all-out, over-the-top approach. Walking around, you can see the design's effect. Visitors are alternately amazed, with gawking expressions, or when they eventually get used to the enormity, relaxed and pleased. When the sun sets, the lights are turned down, and the enclosed spaces glow. The walkways are lit by soft pathway lights. Glowing streamers hang from the distant truss roofs. In the hush, the guests meander among the waterfalls and gently sparkling koi ponds.

The hotel spaces are all artificial, of course. The waterfalls look like they're hurtling over native granite, but it's really textured, formed concrete. The luxury steakhouse isn't perched on a real rock bluff because the cliffs are human-made. Not all of the lush landscaping is alive. Some plants are plastic to reduce the maintenance costs. The flatboats on the river run on a track. Still, even with the obvious Disney touches, you have to admire the enormity and precision of engineering vision and construction that went into building the hotel. The goal was to build one of the most magnificent, whimsical places that could be imagined, and the results don't fall far short of the goal. The infrastructure engineering supporting this work is superb.

It's when you look outside the hotel that you have to wonder. The neighborhood around the hotel is as dreary as the inside is spectacular. The outside features the usual layout of a trashed American landscape. There is the typical life-deadening freeway, a giant shopping mall, endless parking lots, fast-food joints, slow-food joints, and all the other representatives of sprawl. For all of the hotel's excellence, for all of the design's bursting exuberance to create excellent spaces, all of the energy was directed inward, with the result cloistered by the gates. Not an ounce of energy was directed outside, perhaps to build a neighborhood of connected spaces in which the hotel was a part. What if the waterfalls had houses and shops next to them? What if the koi ended up living in real neighborhood ponds? What if the lazy river had public parks on its shores, with playgrounds and basketball courts? What if they tried to build a real town?

Instead, you have another example of the vast American glopscape, the paved-over, sprawled, shopping-malled terrain. At the ASCE National Convention held at the hotel in 2003, you could listen to many presentations on the marvels of the Opryland Hotel design. Technically, the structural design is truly amazing. The architecture, landscaping, and facility operations have resulted in a terrific space. But you would not have heard a lot of discussion, or even recognition, that the underlying system delivering the infrastructure provided a pod of design excellence in a sea of sprawled glop. Perhaps all of us have bought into the unspoken assumption that it is OK for public space to be sited in an infrastructure of crud. So we marvel at the inside places but don't notice what's going on around them.

The Last Game at Foxboro

On a cold, clear day in December, we had tickets for the last regular season game at Foxboro Stadium. This was the biggest football game of the year. It was against the dreaded, first-place Miami Dolphins. If the New England Patriots won, they would continue on to the playoffs and maybe the Super Bowl. However, it was not the last regular season game because it was the end of the season. It was the last game because old, dumpy Foxboro Stadium was to be retired and replaced in 2002 by the spectacular, luxurious Gillette Stadium under construction right next-door.

We parked at one of the small lots a mile north on Route 1 and joined the throngs walking to the stadium. On the sides of the road, fans had set up makeshift camps, with lounge chairs, beer, and barbecues of roasted wieners. These were the tailgate parties. It looked like a scene from a Steinbeck novel, without the Dust Bowl. As we got closer, our path crossed the new stadium construction site. The traffic plan included a new, limited access interchange with Route 1. The superstructure for a three-span, prestressed AASHTO section bridge was in place for Route 1 northbound, but the traffic had yet to be routed over it, and all the sidewalks were ripped up. So, we walked in mud. There were a lot of us now, football pilgrims migrating to the stadium, our feet trudging down Route 1, with the trenchant fragrance of late fall, roasted wieners, and the porta-potties lining the dirt parking lots.

The story of Foxboro Stadium is a cautionary tale for civil engineers. The infrastructure question posed is, "When does an old facility become beloved, or when is it declared a dump?" Not long after Foxboro Stadium was completed, the general consensus was that it was a dump. Even before the big-market professional sport extravagances of the last decade, Foxboro Stadium was a poor cousin to just about every other football stadium in the United States. The structure was basically a double earthen mound, with some metal stands rising on either side of the field. The metal stands didn't have seats in most cases—you sat on benches. Like the planners of the *Titanic* who figured out the number of life boats, the stadium engineers didn't quite calculate the correct number of bathrooms. So, the stadium had an array of strategically placed porta-potties at the entrances, and this was one of its many charms.

The debate about reusing versus rebuilding is something we are increasingly dealing with in this country. Much of our infrastructure was built in the period after World War II, and these structures are approaching the end of their useful lives. The debate has played itself out all over Boston. Consider the three main sports venues. In addition to Foxboro Stadium, the basketball and hockey teams used to play in Boston Garden, and the Red Sox play in Fenway Park. The old Boston Garden was determined to be worn out. It had failed air-conditioning and some seats behind columns, where you literally couldn't see the game without leaning to the side. Although the basketball-playing Celtics, the most successful Boston sport team, won 16 championships there, the Garden was demolished, except for the old wood parquet floor, which was packed up and saved. The replacement arena, built next to the sleek Zakim Bridge, is functional and comfortable but antiseptic and without charm.

On the other hand, the consensus seems to be that Fenway Park, the baseball stadium, is beloved and salvageable. Of the three old stadiums, Fenway Park is in some ways in the worst shape. It requires major structural renovation on account of its age. The small seats are designed for 1920s man—apparently we have gotten bigger and fatter in the ensuing decades. The bathrooms are not much better than Foxboro Stadium, except that there are no porta-potties. But, Fenway Park is an old-time stadium with intimacy, incredible sight lines, and unique quirks, such as the Green Monster, the giant wall in left field that blocks the home-run balls from crashing down on the Massachusetts Turnpike. Instead of demolishing Fenway Park, the discussion centers on how to rebuild and expand it. In many ways, this is a much tougher design and construction job than ripping it down and starting from scratch.

The process of determining when an old structure is rebuilt versus replaced is nonlinear and occasionally irrational. We would do well, as engineers, to bring to the public table our rational, analytic skills. As infrastructure experts, we are the best equipped to lead the discussion. However, dealing with the nonlinear aspects of these debates takes a set of skills and an appreciation for things that we are not used to or trained in.

Back at Foxboro, as the football pilgrims walked through the old stadium gate, we were treated to views of the new stadium. The new Gillette entrance has a large, pedestrian arch bridge. Colorful placards describing the new construction proudly declared that the cost of the new bathrooms is greater than the entire construction cost of the old Foxboro Stadium. These placards were strategically placed at a row of porta-potties.

The game featured tough defense by the Patriots and a few trick plays. The home team ran up a big lead after the quarterback handed off to a running back. Usually the running back would run at this point. That's what the Dolphins' defense thought would happen. The trick was that the quarterback ran instead and was open to receive a pass. It turned out that the running back used to be a quarterback in high school, and he floated a perfect pass for a 23-yard gain. In shock, the Dolphins were pretty much sunk after that. The sun set at Foxboro Stadium in the fourth quarter. The Dolphins tried to rally, but the lead was too big, and the home-team crowd was too hostile. We chanted, "Squish the Fish," although we knew that dolphins are mammals. No one could think of a good rhyme for "mammal." The seconds ticked away, and in the end, the score was much closer than it should have been. The Dolphins had one last play to get back in it—an on-side kick where they could retain possession and drive for a tying touchdown. But their kick failed, and the home team was victorious. The players stayed around for a few minutes to revel, gloat, and shake hands with the fans. Then we departed for the mud, and they turned off the stadium lights.

Postscript

This essay was originally written in January 2002. As we all know, the unthinkable happened shortly after that. The curse of New England sports lifted, and the Patriots pulled off one of the greatest football upsets ever in the Super Bowl. Prior to that was the real last game in

Foxboro Stadium, the amazing snow bowl playoff against the Oakland Raiders, featuring the fumble that wasn't. Redemption doesn't come easily in Massachusetts, land of stone walls, harsh climate, and really bad drivers. But for that shining moment, every ball bounced the right way and every field goal soared through the uprights. That football season is now a memory, other football championships and the improbable World Series have followed, and Foxboro Stadium has been reduced to rubble.

The Zucchini Story

This is a mostly true story that really happened. It has been slightly embellished over the years, and parts that didn't actually happen have been added to improve on reality.

En route to visit a friend in Montreal, we stayed at a bed-and-breakfast in Essex Junction, Vermont. Essex Junction is an important location because it's near Burlington, and it's near the first curved girder bridge design that I worked on. The inn was a beautiful old farmhouse on a hill. The August sun shone brightly, and like everyone else in Vermont, the innkeepers had grown too many zucchinis. After we had stayed overnight, the hosts became comfortable with us, and they asked us a favor—would you please take some zucchinis? The inn was overflowing with zucchinis and the keepers wouldn't take no for an answer. So we said yes.

Mind you, these were not ordinary zucchinis. These were huge vegetables, giant, bulbous zucchinis, of a weight and girth not normally seen in New England. They must have been six feet long—huge green monsters. They were big zucchinis. Really big. The innkeepers didn't just walk out the door with a bag of vegetables. No bag would have been big enough or strong enough to hold these zucchinis. Our hosts had to carry each one individually, and as if carrying a set of barbells or a large family pet. You could see the exertion and strain on their brows as they lugged these giant squashes. Back in the garden, the other vegetables probably quivered in fear at the sight of these gargantuan zucchinis. They were big. Very big.

It was in the days before minivans and SUVs, and we drove a little compact car that couldn't fit many of these zucchinis. We would have had to leave our luggage at the inn to make space. So after some negotiation, we agreed to take two. The innkeepers were a bit disappointed at not having dispatched more of their massive zucchinis, but eventually they were satisfied with the transaction. We did some rearranging in the trunk and stuffed the zucchinis in back, covering them with a blanket so they would be comfortable for the trip. The car tipped back a bit on its rear wheels from the weight of the extremely large zucchinis, but otherwise everything was about the same. We bid adieu to the beautiful inn and made our way across the border to Montreal.

On the way back, driving south in Quebec toward Vermont, we started to wonder. My wife Lauren had heard that it was not lawful to transfer vegetables across international borders. What if it was illegal to import zucchinis? We had gotten away with it once crossing into Canada, but would we be so lucky crossing an international border for the second time? Besides, at that time, crossing the border to Canada seemed more relaxed than the reverse trip. Going into Canada, you waved to the Mounties and that was about it.

As we drove south, the highway started bunching up, an infrastructure tightening of the throat. We were about to leave friendly Quebec for the serious États-Unis. We approached the border barrier structure, which looked like a toll booth but was actually the check station for customs. Cars queued up in different lines. The line I picked moved very slowly. In comparison, cars zipped through on the adjacent lanes. Too late, I realized why. Agents in the adjacent lanes checked cars more casually and quickly moved them back into the United States. My line, however, had a diligent agent. He went over each car inch by inch with a flashlight. He asked some people to open their trunks. Some drivers were asked to leave the main queue and drive over to a separate, special line for a much more scrupulous inspection.

Lauren worried—what was the penalty for carrying illicit zucchinis? All over the gate were signs with red, threatening letters. They probably said things like "Welcome to the United States," but we imagined that they said "We Check For Zucchinis" ("Nous Contrôlons Les Courgettes"). Slowly our car crept up to the gate. An impeccably dressed agent with a large flashlight approached our vehicle and its hidden contraband.

This is a story of the distant past, circa 1985. Decades later, infrastructure design at the borders still deals cautiously and uncertainly with issues more serious than large zucchinis. The customs gate at the Vermont border, along what has been considered the longest and friendliest unguarded border in the world, must now provide space and capacity for all sorts of searches not considered relevant in 1985.

The change is most noticeable and immediate at airports. For example, T.F. Green Airport in Warwick, Rhode Island, was subjected to a complete overhaul in the 1990s. The 1950s-style airport was beautifully redone to become a spacious, comfortable satellite hub to Boston's Logan Airport. The single terminal building features convenient, easy access, soaring rooflines with plenty of light, and comfortable surroundings. Yet what was redesigned only a few years ago has in some ways become outdated. New, massive security lines are required to snake back and forth in unplanned-for queues, covering open areas formerly designated for waiting and greeting. Sometimes the lines are so long that special attendants direct the queues down the stairs to the baggage handling area. As people have adapted to the new inspection requirements, the lines have become shorter and more efficiently managed. But the problem remains that T.F. Green's terminal must deal with new demands and requirements that were not rigorously considered only a few short years before during the redesign.

During that simpler time, we anxiously waited at the Vermont border for the agent to discover our illicit vegetables. Slowly the gentleman walked around our car, gently poking and prodding. He had some questions:

"Where are you folks from?"

"Boston," I said. My wife smiled. She couldn't talk.

"How long were you in Canada?" he asked.

"A few days," I said. "We were visiting a friend in Montreal."

"Montreal. That's a nice city, Montreal. They speak French there. Do you have anything to declare?"

I cleared my throat. Maybe it was time to come clean. Tell the truth now, and they would go easy on us. We had transported giant zucchinis across an international border. The truth would get out one way or another.

"We have some vegetables in the trunk," I said.

"Oh. OK. Welcome back to the U.S." he said, seeming to wave us through. I started to shift to drive. But then the agent motioned me to stop. "Wait," he said.

"Yes, sir," I responded, freezing at the wheel.

"You seem like nice folks," the agent said sternly.

"Thank you," I said. Lauren looked at the ground, tears in her eyes forming at the prospect of our upcoming incarceration. I prepared to drive to the special line for additional inspection.

"You can do me a big favor," the agent said.

"Certainly," I said, a bit confused. "How can we help?"

"You say you have vegetables in the trunk? Would you like some more? I wouldn't normally ask this of strangers, but I'm getting desperate and I'm not sure what to do. I've got these zucchinis here in the booth. I've got way too many of them. The growing was really good this year. I've got to get these off my hands. Can I give you one?"

Moss on the Median

Before Route 3 was widened north of Boston, it offered motorists the illusion of driving through the wilderness. The two northbound and southbound lanes were separated and insulated from their surroundings by wide swaths of woods, both in the median and along the rights-of-way. For most of the 10 miles or so from the Route 128 junction north to Interstate 495, the road seemed to be isolated in its forest and not a ride through suburbia. Come weekends in early October, the old highway was in its glory as the trees started to turn. One could drive at a pleasant clip for miles under a dark blue autumn sky, with orange light sparkling off the shimmering maple trees. There was the promise of pumpkins and crisp Macintosh apples at the country farms along the exits.

This illusion of a drive through the woods was created, in part, by the highway's original method of design and construction. Instead of plowing over the full right-of-way and planting grass, the road was placed and built with comparatively little disruption to the surrounding terrain. This approach was particularly true for the median strip, which remained densely wooded.

Unfortunately, soon enough commuter traffic on Route 3 moved at a slow crawl through the woods rather than a pleasant drive. Traffic studies determined that it was time to widen the four-lane highway to six lanes and improve the substandard interchange layout. Once construction began on the widening project, the first things to go were the trees in the median. Deep gullies were filled in and paved over. The reconstructed highway is well-designed and vastly improved in terms of traffic flow. However, what used to be a drive through the woods has been replaced by a somewhat dreary, characterless slog—Route 3 has been turned into Everyexpressway. The forested median is now an open, grassy space. Whereas the old highway used to fit—in its way—into the landscape, the new version is an open gash that makes as big an impact (and scar) on the terrain as possible.

Limited access highways are a relatively new type of infrastructure, dating back to the 1910s. Some of the earliest highways were "parkways" that were constructed in New York state. These highways had architecturally sculpted viaducts (what we think of as "context-sensitive design" today). They were designed more for leisure travel than for commuting. As the car culture took hold and traffic volume increased, parkways gave way to "expressways." Expressways were the forerunners of the current Interstate highway program, which has standardized and institutionalized the form. Unlike the pastoral parkways, expressways were no-nonsense highways, intended to be fast and efficient and built to provide transportation facilities for easy access and supply of troops during wartime. Only more recently have aesthetics and context-sensitive design, where the facilities are understood and even designed to be part of the surroundings, been considered. As part of these recent improvements, expressway bridges now have road and place name markers attached, in an effort to help drivers feel connected to the landscape through which they are driving.

Parkways became more widespread thanks to Robert Moses, the late commissioner of the New York Triborough Bridge and Tunnel Authority. Under his leadership, parkways and many aspects of the suburban form were developed around New York City, particularly on Long Island.* Parkways were designed to be not just limited access highways, but an aesthetic experience. Therefore, the rights-of-way featured trees and fields, and the overpasses had stone veneer and architectural details. Parkways were not originally conceived as massive people movers, but as roads for pleasure drivers. Some unsavory aspects of the original parkway designs included such features as deliberately low overpasses intended to keep buses and mass-transit-riding city dwellers from using the parkways and thus invading suburbia.

Today, on Long Island, you can drive on two types of freeways: the typical, brutal expressway like the Long Island Expressway or one of many parkways such as the Northern Parkway. The parkways' original

---

*Caro, Robert. (1975). *The Power Broker: Robert Moses and the Fall of New York,* Vintage, New York.

Sunday drive function has long been superseded by the hordes of commuters that motor back and forth each day. The Long Island parkways still have wooded rights-of-way, and the viaducts are still sheathed in attractive stone. But these freeways have been rebuilt over the years for the purpose of moving traffic, with the addition of lanes and modern geometrics at their interchanges.

Traveling north from New York City, drivers have an interesting choice in Connecticut between the Merritt Parkway and Interstate 95 (originally the Connecticut Turnpike). Merritt Parkway is an historic highway with neoclassical and modernistic style bridges festooned with all sorts of ornamentation. Except for the crushing traffic, a jaunt on Merritt Parkway is pleasant and invigorating, with long stretches of woods and a canopy of trees overhead. Interstate 95, on the other hand, is a particularly ugly, placeless highway that takes no prisoners as it makes its Sherman's March across the landscape to New Haven.

With time, Interstate 95 has softened, and the Merritt Parkway has hardened. Interstate 95 has had reconstruction and landscaping improvements so that it is not the scar it used to be. Much of the reconstruction has been nicely done, with aesthetically shaped, attractive viaducts. On the other hand, the Merritt Parkway has been subjected over the years to many ill-advised reconstruction projects to improve traffic flow while ignoring the historic design of the bridges and woodsy layout of the terrain. Several miles of the parkway are protected by aluminum crash barriers that have not weathered well and do not fit the parkway theme.

Recently, some groups have organized against the Merritt Parkway rebuilding. In June 2005, the Merritt Parkway Conservancy and the National Trust for Historic Preservation, together with the Norwalk Land Trust, the Norwalk Preservation Trust, and the Norwalk River Watershed Association, filed a lawsuit against the Federal Highway Administration, seeking to downsize the massive interchange project at Route 7 and Main Avenue in Norwalk, Connecticut. The project was originally designed assuming that Route 7 would become a full-capacity north–south expressway. But Route 7, the expressway version, was never completely built, so the plaintiffs argue that the project no longer makes sense.

If you drive down the new Interstate 495 between Norton and Mansfield, Massachusetts, you will encounter a 20-mile straight stretch of six-lane highway with a very wide grassy median. It's easy to imagine what the same space would look like with woods preserved in the median. One argument against a forested median is that it would be undesirable and unsafe to have deer and other large animals inhabiting that space and darting into traffic. But animals have found a way of adapting to suburbia anyway, and a counterargument maintains that a boring, featureless road can result in more accidents. In terms of land usage, the Interstate 495 right-of-way features a lot of wasted, useless space. If the wide median cannot be filled with trees, a better approach would be a layout similar to the Massachusetts Turnpike or the first nine miles of Interstate 95 in Rhode Island, east of the Connecticut border, where there is no wide median and for safety the lanes of traffic in each direction are separated by a tall, concrete Jersey-shape safety barrier.

At some point in the future, all of what is currently unused land will be used. Some estimates say that eastern Massachusetts will be built out by 2050.* Wooded highway median strips may start to fulfill a role never intended by designers: miniature habitat preservation in areas where the natural habitat has largely been plowed under. Some wooded medians are particularly wide, such as Route 128 in Dedham, and Interstate 95 in Connecticut just west of the Rhode Island border. These unused tracts feature acres of preserved woods that will likely never be developed (although in the case of Route 128, a use was found for part of the land in placement of a new prison). The woodlands are peculiarly sited for preservation, being surrounded by streams of vehicles on all sides. At some future date, when all initial land use development is done, we will drive on some stretches of freeways that still appear to be carved out of the forest and the roads will remind us of a time when the land was in a wilder state.

*Kirshen, Paul, Ruth, M., Anderson, W., and Lakshmanan, T.R. (2004). "Infrastructure Systems, Services and Climate Change: Integrated Impacts and Response Strategies for the Boston Metropolitan Area," report prepared under U.S. EPA Grant R.827450-01. http:// www.tufts.edu/tie/climb/CLIMBFV1-8_10pdf.pdf (accessed July 13, 2006).

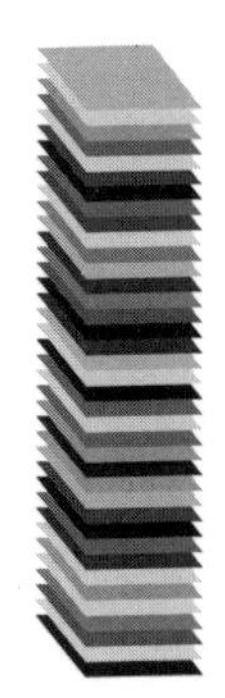

First Class

Buried deep in my wallet was a form for an upgrade to First Class. I was scheduled to fly back from Seattle to Boston in Coach. I wasn't looking forward to this flight, because the earlier trip from Boston to Seattle hadn't been very pleasant. The plane was packed, and for five hours I was stuck in a tiny center seat between the aisle and window. In this arrangement, you have to frequently move your legs, only to have your back or some other body part lose circulation. The challenge is to squirm around so that the blood flows enough to avoid later amputation of dead appendages.

When I pulled out my wallet at the gate, the old crumpled upgrade form fell out. Once I realized what it was, I sheepishly handed the form to the agent. It was like at the end of *The Wizard of Oz,* where the Wizard bellows at the Scarecrow behind his magic curtain: "What? You want Brains???" The agent smiled at me and started typing in his magic computer. After a few fretful minutes, I was instructed to be seated, and after appropriate deliberations, the agent would render judgment on my request. Apparently the stars were in line that day, because I was high enough up on the list, and there was one seat left. I was about five feet away from the desk, but just for the public spectacle of it, the agent called my name on the terminal loudspeaker. He had some good news. I was going to fly First Class the whole way back to Boston.

They started loading the plane with those pompous boarding announcements: First Class, special people, and the indigent could board first, and then the riffraff would be called by their seat rows. We privileged passengers in First Class raised our noses in a snooty salute and sauntered down the Jetway. In the first-class cabin, there was calm music.

We seated our corpulent butts in the plush, oversized lounge chairs. In First Class, your legs don't touch the seat in front of you. Our staff served chilled beverages, not in plastic cups, mind you, but in real glass glasses, with the executive logo of the airline embossed on the side. Once we were in the air, it was time for breakfast. We were given a menu with a choice of several selections. The fruit was ripe and succulent. The waffles were fresh and fragrant. The coffee was as bad as always, with that tinny jet plane taste, but it was delightfully served in real china and not a paper cup. In Coach, the riffraff was served the bad coffee in paper cups.

Occasionally, the real world intruded. We could hear the moans from steerage, and sometimes the odor of the unwashed wafted into our sanctuary in the front of the plane. But for the most part, we were First Class, with a secluded, private world to ourselves. We dined in luxurious comfort thousands of feet above the clouds.

In that pleasant aerie, I thought about what it meant to be in First Class. Americans have had an uncomfortable, somewhat contradictory relationship with the concept. Our capitalist, business culture leads us to want to achieve the best, to go First Class. Yet our democratic traditions mean that everyone is equal in the pursuit of happiness. Therefore, First Class is not a divinely granted state, or something bestowed on the fortunate by an artificial class structure. It is a condition that you earn, and everyone is free to try to earn it. So it is simultaneously exclusive and open to all. We strive for First Class, but at the same time we are ill at ease with the intimations of privilege and exclusivity.

Our discomfort with the ideals and trappings of First Class is epitomized by the story of the *Titanic*. Not only was this ship the biggest cruise liner built, but also it featured accommodations of great luxury and spectacle, at least for the lucky passengers in First Class. After the iceberg collision, when the ship started to sink, first-class passengers had the best access to the lifeboats. Many more passengers in second-class and steerage perished because there were no slots left on the boats. By American tradition, this was an abuse of the idea of First Class. Regardless of level of service, everyone should have had equal opportunity to survive. We want to strive for the best, but at the same time we want equal access.

Some have commented that our democratic traditions have led to a downgrading of transportation service. For example, nowadays, flying is relatively inexpensive, so almost everyone can fly. But, the old mystique of an exotic, mysterious airport terminal has been replaced with that of a bus depot, with the recently added ambience of a police station. To go First Class in the past was to travel in great luxury and privilege. Today, traveling in First Class is more like "adequate" class, where you get to go without the physical discomfort of Coach.

You can see the contradiction of First Class at work in our engineering, in the tension between the ideal striving to go First Class and the reality of getting the job done. The conflict is optimizing a particular engineering task versus fitting all of the tasks together so that the whole is successful. Sometimes you have to fly Coach to balance all the demands. But, both Coach and First Class ultimately arrive at the destination and result in a quality resolution.

Soon it was time for the plane to land. Using silver tongs, the staff distributed refreshing warm towelettes. The wheels touched down, and we were back in Boston. As I left the jet, I looked back one more time at the First Class cabin. The staff smiled and wished me a safe journey. Then I rejoined the bleating masses herding down the Jetway. I was no longer First Class, but I was free to pursue that status again.

After All, It's a Small World

My son Dan and I decided to conduct a little experiment. We were staying near Hyde Park in London. This is the beautiful park where Mary Poppins sang to animated finches. The park has wide, expansive fields, so we thought it would be a good idea to demonstrate American sports to the Brits. After seeing us playing real sports, our hope was that they would stop getting all worked up about soccer, cricket, and other such nonsense.

We started with something simple—throwing a Frisbee. We hiked over to a field next to the lake. It was an uncharacteristically warm and sunny day in London, and many natives were out in the park playing soccer and drinking tea. We tossed the Frisbee back and forth for several minutes, executing some perfect tosses. Unfortunately, the appearance of the Frisbee did not seem to dazzle the natives as we had hoped. They hardly gave us a second look, continuing their soccer scrimmages.

OK, that didn't work. We tried a football toss next (a real football, not a soccer ball). Surely the sight of the football, so bold, so American, would shock them to their senses and bring them into the real world of sport. We tossed some spirals and ran a few plays. Dan reminded me that I didn't really know how to toss a spiral. We tried playing a game of one-on-one, using landmarks in the park for goal posts. This game lasted only so long, since one of the players was significantly faster and more agile than the other. But it was a spectacle of excellent football. Use your imagination, and you could see throngs of spectators in the open-air stadium in November, screaming obscenities in an understandable accent of English. The frost was long on the pumpkin, and the tailgaters were out in full force, drinking beer, roasting knockwurst, and in general appearing like refugees from *The Grapes of Wrath*.

By now, surely the Brits were ready to drop David Beckham and adore Tom Brady. But our football display didn't have its intended effect, either. The natives went about their business, playing soccer and hardly giving us notice. We were foreigners playing a strange sport in Hyde Park. What was the difference here? As far as we could tell, the physical appearance of Great Britain was not much different from the States. As an expression of culture, much of the infrastructure was the same and identifiable. The buildings were older and more festooned than American structures, as befits a European city, but the dimensions and particulars were about the same. The structures had doors and windows and other accoutrements. Likewise, the transportation infrastructure was easily identifiable to us, with the possible, sad exception of drivers using the wrong side of the road. Traffic lights, road markings, and dimensions were about the same as what we were used to. There was something called a "humped zebra crossing," which despite its unfortunate name was really just a raised crosswalk with stripes. British engineers often eschewed T- and cross-intersections in favor of rotaries, which they named "roundabouts." There were a lot of them.

But all in all, the built form looked a lot like what we were used to. I thought about how a culture expressed itself in terms of infrastructure. In the past, foreign places were really foreign and exotic, and the differences were clear in their physical layout. With the world shrinking and worldwide communications almost instantaneous, differences among places have narrowed. Is there a developing human standard for how infrastructure should look and function, sort of like an imposed operating system for all computers?

Since the United States has led the way in achieving a high standard of living for its residents, it has also apparently led the way in imposing a type of standard physical expression of the built form. For better or worse, limited access highways, shopping malls, and the like were developed in the States. It seems that much of the rest of the world aspires to American levels of physical well-being and comfort and has increasingly adopted many of these forms. If this is true, soon every place will look like southern California. For example, the British equivalent of interstate highways, called motorways, are very familiar to American drivers, with that unfortunate exception of misplaced driving sides. The British have improved upon the basic form to an extent, with excellent geometrics and a superb if overbearing, nanny-like system of electronic warning signs every mile or so. But the highways' basic form is instantly recognized by Americans, not as something from the planet Mars, but as a limited access highway. The signs are even in English!

If everything else is increasingly homogenized and the same, then perhaps whatever differences remain between cultures are to be held on to. Maybe this was why our display of proper sports had little impact on the natives. London was overrun with Starbucks. They had enough American things. They were happy to play cricket and soccer.

The next part of our experiment didn't actually happen. We decided to invent a sport and see if that would shake them out of their slumber. We took a red dodgeball and fastened to it the halves of six cucumbers, one at each pole and four around the equator. Grasping the ball (which we called the "flingallee"), we tossed it back and forth in a particular pattern, not too far and not too close. After a really good spin from the flingallee, we bent down on one knee, raised our arms to the sun and cried out in joy, "Huzzah! Huzzah!" A Brit walked by with a pensive, curious expression and said, "Jolly good show, chaps. Spot of tea with that?" Then he stopped what he was doing, forgot how to play cricket, and joined us.

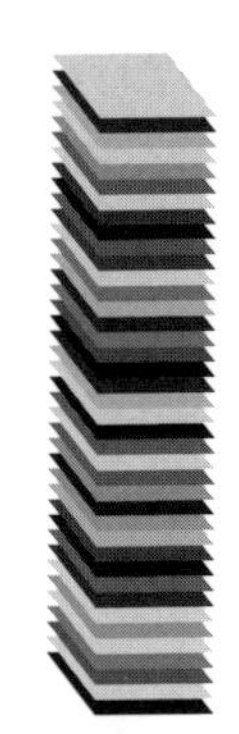

What Happened to Nantucket?

I biked with my son out to Madaket on the western edge of Nantucket Island, about a six-mile ride from the center of the old whaling village. The weather was cloudy and cool, with a type of gloominess that Nantucket excels at. On the beach looking west over the water, you could see every shade of gray imaginable. The mottled sky blended into the gently swelling water. There wasn't a foghorn blaring in the distance, but you could imagine one. Probably out on the horizon floated a lost, 17th-century ghost ship.

The beaches of Nantucket are among the most beautiful in New England, and this is a large part of the island's appeal. There is nothing particularly unique about a beach in summer, but on Nantucket, the beach and seascape express a special type of moodiness. At Madaket, the bluish-gray swells gently lapped the sand, rubbing against the shells and driftwood. Along with the imaginary foghorn and the real sea gulls, it was a great place for reflection. This essay is a reflection about infrastructure.

In addition to the beach and ocean, Nantucket has an amazing old fishing village. The village is situated on a gentle hill sloping down to its well-protected harbor. The village is a collection of preserved, historic buildings and cobblestone streets. In the 19th century, Nantucket was the whaling capital of the world. For a brief period before the discovery and use of petroleum, Nantucket supplied much of the world's whale oil. This substance was used to light the lamps all over the world, so the island was very wealthy. The whaling captains built spacious, grand mansions with widow's walks and fancy turrets.

What is impressive about the village is not just that it's old and preserved, but that the buildings and streets are well-designed and attractive. It's not only the individual structures, but the way they come together to form public space and the village as a whole. Each street is a human-scaled outdoor room with trees and buildings placed at just the right proportions to create inviting spaces. Walking around this village, whether on the bustling main street or in the back alleys and side streets, is enjoyable and exciting. The construction of private structures in Nantucket Village led to the creation of excellent public spaces. The sum ends up being greater than the constituent parts.

Note that Nantucket is no longer wealthy because of whales, but because of tourism. A lot of money has gone into restoring and maintaining the old village. Probably the village spaces would be a lot less appealing if they were run-down and poorly maintained. That intimate public street space would feel oppressive and threatening if it was physically decaying. Also, the island is a vacation destination, populated by hundreds of carefree, relaxed, happy people. Being surrounded by these people goes a long way toward making a public place enjoyable. Yet, even with these considerations, it's important to understand how the infrastructure design and layout create such a worthwhile place. Compare downtown Nantucket village to your local strip mall. The arrangement of stores, roads, and infrastructure in the strip mall creates spaces where the sum is less than the constituent parts. The message implied by the design of the mall parking lot is that you need to get out of it and into the store as quickly as possible. In fact, this is exactly what is intended and expected.

It's not a question of old and new, either, but of infrastructure design. The old seems quaint and desirable. The new seems unpleasant. However, it's not because one is old and the other is new, but because of the scale of the human-made environment and the way the structures and facilities make more out of the individual parts rather than less. New infrastructure can be well designed and built. Old infrastructure can be poorly designed and built, although we tend not to see this as much, because badly designed old infrastructure ends up demolished and replaced.

The forces responsible for your local strip mall and office parks are also at work on Nantucket. Between the mournful seashore and the village are stretches of moors and scrub pine woods. In the past, there was a distinct edge between the village and the countryside. That edge has now been blunted, suburban Nantucket style. Large summer homes have been dropped down on plots of land. The homes, while individually beautiful, tend to have no spatial relationship to each other. In fact, they seem to have fallen from the sky onto the landscape. Instead of creating a beautiful village like the old whaling captains' mansions, these new mansions create an uninviting plopscape of private spaces on what used to be the moors and woods of the hinterlands. You can see this effect most distinctly by biking down the old Polpis Road on the island's northeast side. The hulking summer homes seem to be a chic, gray-shingled parody of housing, and taken as a whole, the landscape is ugly.

The residents of Nantucket have been conducting an intense debate about development. Even though it's an island, there are an awful lot of cars in the summer. The vehicles congregate in congested, fume-laden traffic jams. While the village is dense and walkable, and you can bike to the beaches, it's really necessary to have a car (or two or three) to access the distant summer homes scattered around the island. To address the problem, the town introduced a mass transit system of sorts with several bus routes. Yet the root issue, that of land use and private rights versus public impacts, remains to be addressed in any meaningful way. In that sense, Nantucket is a microcosm of the land use trends and issues faced all over the United States.

You can still experience and imagine Nantucket as it was. To the island's credit, much of the moors and woods have been preserved, particularly along the Milestone Road in the island center. But the 21st-century version of Nantucket seems to have been degraded from its past beauty. The beaches and village are still wonderful, but you need to hold your nose at much of what has recently been built in between.

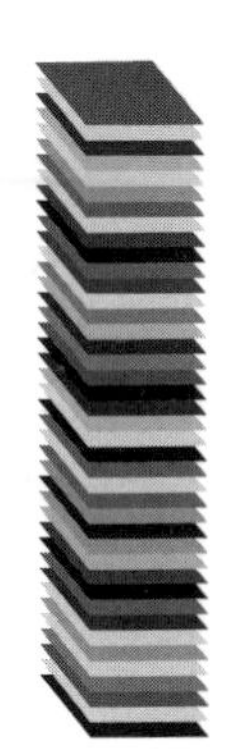

The Way Things Are

After many years of frustration, my local library gave up. Keeping with a practice begun since the dawn of libraries, they had placed reusable, blue due-date insert cards in the book jacket sleeves when a book was checked out. The librarian would stamp the due date on the blue card. The problem was that readers were using the blue cards as bookmarks, instead of keeping them in the pouch where they belonged. When used as bookmarks, the blue cards didn't perform very well. They would get bent, beaten up, and gnashed, and thus would fail to serve the function of being reusable. The library's replacement, a clever idea, was to use yellow sticky notes. The librarian would stick the yellow sticky in the back jacket of the book you borrowed and stamp the due date on it. Yellow stickies are cheap and don't need to be reused, so apparently the problem was solved.

A few days ago, I returned a book I had borrowed. I had used the yellow sticky as a bookmark. The librarian glared at me. She said, "You know, you really shouldn't use these as bookmarks. We have bookmarks for that." She handed me a bookmark. I had that feeling of being back in the fifth grade facing the teacher's reprimand for whatever. I meekly replied, "Yes Ma'am," and carefully placed the new bookmark where it was supposed to go. The librarian smiled approvingly.

Afterwards, I fumed a little bit about how I couldn't come up with a better response at that moment. My written sarcasm is pretty good, but in that situation I had the deer-in-headlights look. After I finished fuming, I thought about what motivated the librarian. After all, they had ditched the blue insert cards in favor of easily replaced yellow stickies so that reading slobs like me could do the obvious and use them as bookmarks.

Maybe there wasn't a policy written about this, but it was pretty clear that it was allowed. It's not as if the library had a desperate need to reuse the yellow stickies.

I came to the conclusion that the librarian was set in her ways. For years, she lectured hapless readers about the need to use designated bookmarks for bookmarks. This was probably built into librarian training. She understood that bookmarks were to be used for marking the page, and the blue inserts, which had now morphed into yellow stickies, were to be used for the due-date reminder. Even if the policy changed, this was the way things were supposed to be done. It reminded me of a scene in the movie *Babe,* which has talking pigs, dogs, and other farmyard animals. The movie is a kind of parable about the way things are. The dogs are supposed to guard the sheep, and the pigs are supposed to get eaten at Christmas. The title character, a pig, questions his fate, and his adopted mother, who is a dog, councils him that this is the way things are. Since *Babe* is a children's movie, the pig is not eventually eaten, but there is a duck who does meet her fate with l'orange sauce (fortunately, this scene is tastefully filmed).

It seems that it is human nature for us to define ways of doing things and then live our lives within these boundaries. It may be irrational and counterproductive, but it's "the way things are" and that's what we do. The boundaries can give comfort and order to situations that are inherently uncomfortable and disordered. Personally, I'm often guilty of this type of behavior. I go jogging outside my house in one direction and only one direction: clockwise. When my son was old enough to go running with me, he wanted to know why we couldn't go counterclockwise. I had to explain to him the clear, understandable, obvious reason: "Because."

In engineering, we can think of all sorts of examples of the way things are. For example, in traditional design-bid-build contracts, we have a certain expectation about how the different players are supposed to perform. The designers are thought to be the virtuous guardians of quality, and the contractors are assumed to be more focused on cost and schedule. Whether this is true or not, it is the way things are, and engineers are used to functioning in the design, bid, and construction phases with this underlying assumption.

When change comes along, reacting to it can be difficult. In a chapter of ASCE's *Quality in the Constructed Project,** all sorts of new and bewildering project delivery methods are described in addition to the comfortable, familiar design-bid-build approach. The methods include design-build, design-build-operate-maintain, and many others. Each of these methods implies different relationships among the players. But, as designers, do we hold onto certain ways of acting because of expectations that this is the way things are? Using the design-build process, expectations for satisfying requirements of quality fall more directly with the contractor. How could this be, since considering the way things are, one might expect that quality would be in its usual order of importance behind project cost and schedule.

But, the way things are has become the way things were, and we are forced to change to adapt to the methods. Change is difficult but possible. In the case of our business practice, it's a requirement or we won't be in business (this outcome would also be a change to the condition of being in business).

Eventually, I went jogging with my son counterclockwise. It felt strange at first and incorrect, but I got used to it. With that change under my belt, I became daring. I was ready to push the envelope. That night I tossed the new bookmark and used the yellow sticky.

*American Society of Civil Engineers. (2000). *Quality in the Constructed Project, 2nd ed.,* ASCE Manuals and Reports on Engineering Practice No. 73, ASCE, Reston, VA.

The Forest and the Trees

When I was a younger engineer, I lived among the trees. As a bridge designer, I was familiar with all the details of AASHTO, ACI, AREA, and other codes. My structural analyses were good, my details were detailed. At this time, the computer revolution was starting to sweep through the civil engineering world. I rode on the crest of the waves. I was up to speed in the latest nuts and bolts of all the programs. I could make spreadsheets sing, with automated macros that did all the analysis and design, well, automatically.

I was told by my seniors that engineering is not just a science but an art. They said that not everything was a precise, quantifiable, analytical problem. There were nuances, shades of gray in every situation. There was more than one way to solve each problem. But of course, I was convinced that the seniors were wrong. They couldn't see the trees, but I could, living next to them each working day.

As I got older, I started to move away from the trees and live in the forest. I began to learn that much of the bunk that the seniors spouted wasn't such bunk after all. I began to understand the bigger picture. In any endeavor where human beings are involved, there is imperfection. Civil engineers in particular butt up against all sorts of conflicting motivations. In any one project, the goal is not the "optimal" solution of one constraint, but a solution that satisfies many different constraints. Sometimes the constraints aren't easy to quantify by numbers.

As the old saying goes, "he couldn't see the forest for the trees." Or perhaps, taking into account the progression of time, "he couldn't see any of the trees in the forest." Considering my career path as a practicing engineer, it got me thinking: is it possible to live in the forest and still appreciate the trees? In terms of engineering information and responsibilities, the forest is getting a whole lot bigger, and the trees are more dense. For example, what was once a relatively simple seismic structural code has gotten progressively more complicated. You used to be able to do things like pseudostatic loading and calculation of simple base shears. Not any more. With each new earthquake, we learn new things about how structures behave. This trend of increasing information is true for every engineering discipline. To use the forest analogy, more trees sprout, and the canopy thickens.

I think that our engineering education system can do a better job to assist more senior engineers in understanding the details. Engineering students and young engineers are expected to participate in all classes focusing on analytical skills and the details (the trees). Older engineers are expected to master "the bigger picture" (the forest). Classes for them focus on management, human resources, etc. I wish there were more classes like "Intensive Bridge Analysis for Managing Engineers Who Don't Need to Know How To Do It, But It Would Be Nice If They Did." Such classes need to come with an expectation that senior engineers should participate in this type of thing. The movement toward continuing education credit for professional registration seems related to this, but it seems unfocused so far, and possibly of more benefit to a bureaucracy than to practicing engineers.

I suppose it's impossible to know everything about everything (which is what I really want). The trick, then, is to be able to find the right balance between the details and the bigger picture—to be able to live in the forest and appreciate the trees.

Mass MoCA and the Hoosac Tunnel

Mass MoCA is not an espresso drink, but an elaborate museum in North Adams, Massachusetts. The museum displays works of modern art and "installations." To a rational engineer such as myself, a lot of the artwork is hard to appreciate. The material can be confusing and unapproachable. When I visited, the museum was displaying something about risk and the cosmos of the universe. You threw some weird dice at a sort-of craps table and then wandered around the hall to match your playing card. I had a Joker with a symbol representing eternity, or the loss of lunch, or something like that. There were many paintings on the wall. I think they were paintings. They had blobs of color arranged (or flung) in random patterns.

Next to this exhibit was a vast space filled with crumpled white paper. New sheets of paper gently floated down from the high ceiling at timed intervals. While this was happening, eerie voices cried out from tethered hanging speakers that raised and lowered overhead. Children ran around in this space. This installation resembled an untidy bedroom, only much, much bigger with a lot more stuff on the floor.

At the risk of seeming gauche, I admit that I didn't understand the point of most of the exhibits. One that I did get (in fact the only one I got) showed video clips of a taxi wandering around the town. The taxi had a message board on top that had been programmed with context-appropriate messages automatically matched to a GIS. For example, if the taxi drove by McDonald's, the message might be something like, "Eat healthier food." Now this was a clever exhibit, and it even had a computer database and engineering things. But overall, most of the exhibits struck me as peculiar. At the top of the strangeness list was an exhibit I can't even describe completely. Suffice it to say that this exhibit displayed a male mannequin wearing a gold foil suit that had a dynamic feature.

I may not be smart enough to understand modern art, but I do like irony, and I thought that Mass MoCA was plenty ironic. The museum occupied a beautifully restored old factory complex. The complex featured a campus of 19th-century brick buildings, redone in industrial chic for the museum. Along with the strange museum was a strange restaurant and strange store. The idea, apparently, was to develop a center of avant-garde strangeness. A lawyer's office was located in one of the buildings—not strange per se, but the lawyers probably liked being associated with the edginess of the complex.

The Mass MoCA creators had succeeded admirably in establishing the museum after years of battling adversity. The museum complex has become very popular and a destination attraction, no easy thing in the fiercely competitive Berkshires with such stellar cultural attractions as Tanglewood and the beautiful Williamstown art museums. The facility became a catalyst for redevelopment of the whole city, which, along with Pittsfield, had been a poor stepchild to the rest of the classy Berkshires. The Mass MoCA museum was a roaring success and helped to put North Adams back on the map. The downtown was thriving when I visited, with a mix of eclectic and practical stores on the busy, attractive Main Street. North Adams even had a real mocha bistro, something unimaginable in the dowdy, hardscrabble little city of the past.

The irony comes when you think about the previous uses of the factory complex. Back in the 19th century when the factory was a factory, workers made widgets of some sort. I'm not sure what they made, but it was clearly something tangible, corporeal, and useful. In the days of competitive American manufacturing, the understanding of what you did with your time and the resulting product was different than today at Mass MoCA. Workers trudged off to the mill in a pre-post-North Adams factory setting—we're talking Flintstones here. In the 1800s, the frontier at the edge of the wilderness was much closer. Not far from the mills that became Mass MoCA are the portals of the Hoosac Tunnel. In fact, the tunnel's west portal opens in North Adams, about four miles from the museum. The tunnel was constructed to open a more direct connection from eastern Massachusetts to the vast frontier lands beyond the mountains.

The Hoosac Tunnel is a great civil engineering achievement. It was completed in 1875, and in its day, it was the longest rock tunnel in the world, extending 4.75 miles below the Hoosac mountain range. The tunnel opened a clear railway route to the west. It required 20 years to build and featured the first use of nitroglycerin. The tunnel was built at a time when the connection between things and what you did to get them was much more direct and understood than it is today. Freight traffic still uses the tunnel, but its importance has been eclipsed in the roar of semis barreling down the Mass Turnpike to the south.

In today's factories in North Adams, the frippery of Mass MoCA revitalizes and drives the economy forward. The infrastructure comes back to life, and the area is reborn. But I wonder if this is a real rebirth. The art installations at Mass MoCA, while maybe entertaining if you are wise enough to understand them, don't really build or accomplish anything. This is a silly argument, I know—there is a place for pure abstract (and in this case, completely disconnected) thought in the scheme of creation. All the same, my son is also good at making a mess on the floor, and this is not enough to drive an economy and revitalize a city. Maybe I'm too much the engineer, looking for more direct cause and effect. Maybe I should accept that food just arrives at the supermarket and not wonder about how it got there, and what I'm doing to earn the right to eat it. But in North Adams, there are ghosts in the factory walls and in the damp musty darkness of the old tunnel, looking on and wondering as well.

The Trail Ridge Road

On the Trail Ridge Road in Colorado's Rocky Mountain National Park, you can drive high into the mountains above elevation 10,000 feet. The road is only open in the summer, from June to mid-October. During the rest of the year, it's buried under many feet of snow. Above the tree line, you enter another world. The saw-toothed peaks are windswept and covered by a white blanket. You drive along the edges of soaring vistas, with sweeping views of the sky and mountains in all directions. The snow and ice sparkle in the brilliant high-altitude sunlight. The place seems unreal and unconfined.

At a point on the road called "Rock Cut," there is a turnoff for a trailhead. The trail is a half-mile hike along a lonely, winding path, through meadows of tundra and frost. Everywhere, you hear the howl of the wind. Summer wildflowers struggle to remain rooted in the tundra and not be blown to oblivion. Birds fly into the wind and are pushed backward, like salmon trying to swim upstream against a too-strong current. At the trail's end is a kind of natural rock kiosk, with a panoramic, 360-degree view of the mountain spectacle. There, in a windbreak formed of slabs, a metal plaque is fastened to the rock wall. The plaque is a human-made artifact that seems strangely out of place in this forsaken mountain wilderness. The plaque honors Roger Wolcott Toll, superintendent of Rocky Mountain National Park in the 1930s and the chief engineer of the Trail Ridge Road.

The road is an impressive engineering accomplishment. It snakes its way from the valley floor up to the mountain crests, with modest grades and minimal cuts and fills. The achievement is more impressive when you remember that the grading, calculations, and layout were all done manually, without COGO or AutoCAD. The road was built, in part, to give more people the chance to experience the amazing mountain vistas. It is one of the world's great alpine highways.

Some historical background from the National Park Service:

> Construction on Trail Ridge Road began in September, 1929 and was completed to Fall River Pass July, 1932. Trail Ridge was built to counter deficiencies of Fall River Road. The historic, gravel route was too narrow for the increasing numbers of vehicles. Frequent snowslides, deep snow, and limited scenic views also plagued the route. The maximum grade on Trail Ridge does not exceed 7%. Eight miles of the road is above 11,000 feet in elevation. Two different contractors were hired to complete different sections of the road. The first section completed, 17.2 miles, was Deer Ridge (8,937′) to Fall River Pass (11,794′). Later, in the early 40's, this section was paved.
>
> During road construction, workers had only about 4 months of the year (mid-June to mid-October) to work. The presence of permafrost required that careful attention be paid to construction to avoid permanent quagmires. Planning efforts sought to reduce scarring on the surrounding landscape. Natural construction debris was removed. Log and rock dikes were constructed to minimize scarring and scattering of rock blasting debris. Extra surface rocks were placed lichen-side up. Tundra sod was salvaged and carefully placed on road banks. Rock projections were kept as scenic "window frames" instead of being blasted away. Rocks matching the surrounding land were used for rock walls. (http://www.nps.gov/romo/visit/weather/history.html)

In the era when the road was built, attitudes about nature and the wilderness were much different than they are today. These attitudes, in turn, colored expectations about civil engineers' responsibilities. In the 1930s, the frontier was still menacing and inhospitable, something to be developed and conquered for humanity's benefit. Civil engineers were the conquerors. Nature was something to be tamed and reshaped for humanity's comfort, not an environment to coexist with. That the process of taming could result in the end of the frontier was not a widely understood consideration, although ideas like preservation and establishment of national parks were being discussed. The dominant theme was that engineers presided over the process of creating civilization out of the wilderness, and it was good.

In the 21st century, the Trail Ridge Road is, in a sense, an anachronism. The idea of the land as wilderness to be conquered is a dated concept. Today, there is more widespread understanding that we are part of the natural environment and not separate from it, even if this understanding does not yet greatly impact land use. Not far from the spectacle of Rocky Mountain Park is the Front Range. This is the edge of the Rocky Mountains, where the Great Plains abruptly end at a startling and surreal wall of mountains. The Front Range today is not much like Native Americans and the first European settlers would have seen it. The city of Denver has exploded north along the range, in an exurban sprawl that has enveloped Boulder and continues north 80 miles to Fort Collins and beyond. The sprawl features the usual hodgepodge of unrelated office parks, shopping centers, and housing developments, threaded together by overstuffed freeways and arterial highways. The sprawl isn't much different than the glopscapes found elsewhere in the United States. The sprawl is only made notable, perhaps, by the backdrop of the wall of mountains, at least on those days when you can see the range through the auto exhaust smog. Not too far north is the Wyoming border. Wyoming is one of the wildest U.S. states, and now it is about to be surburbanized. There are no obvious barriers to the upcoming paving and Wal-Marting of the Wild West. Instead of lassoing the dogies, you will catch a burger at McDonald's.

A new set of skills and attitudes is called for to tame this new version of the Not-So-Wild West, skills and attitudes much different from the approach used for building the Trail Ridge Road. This new and difficult debate has just begun, and it is one that we civil engineers must participate in. It was easy for the public at large to understand the concept of the taming of the wilderness and our heroic engineering role in doing so. The new job of taming land use planning, and of designing a built environment where the pieces fit together, involves more ambiguity and conflict than the simple black and white of the old job of the subjugation of nature. To an extent, it is up to us whether we engineers will be seen as heroes or villains in this new drama.

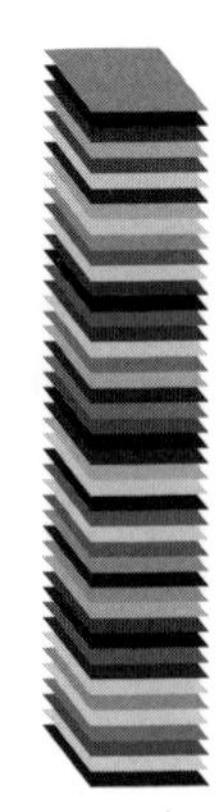

Encino Engineer

I saw *Apollo 13,* one of the great engineering movies of all time (not that there are that many to choose from). For those unfamiliar with the (real life) story, an oxygen tank explodes en route to the moon, and the engineers and astronauts frantically improvise solutions to get the spaceship safely back to earth.

Two scenes stick out in my mind.

In one scene, a crucial calculation must be made by Mission Control engineers in Houston. The calculation must be done and checked in seconds, or all of the astronauts on the crippled spaceship could die. Of course, this was in the days before PCs on every desk top. The engineers hover around the table with their slide rules, furiously calculating as the precious seconds tick away. Finally, one engineer announces the results, and the checker confirms that it's accurate. As the slide rules cool, the calculations are radioed to the space ship, an adjustment in course is made, and the crew is saved.

As a structural engineer, I've prepared many sets of calculations. None were completed under life-or-death conditions. It was sort of fun to watch the do-or-die circumstances of this set. The engineers were portrayed as balding and sort of nerdish, but that didn't matter. They were heroes.

In another scene, the crew has moved to the lunar landing module. The main crew compartment must be shut down, and the lunar landing module is to act as a lifeboat for the perilous return journey back to earth. The module is required to perform in all sorts of ways for which it wasn't designed. At one point, a representative of the company that built the module is speaking to the head of Mission Control. As the demands on the module's performance are piled on, the representative complains that he can't be sure it will work. He can't be sure that three people can be transported in the craft that was only designed for two. He can't be sure that the lunar landing module rockets can be used to correct the course of the damaged spaceship. It wasn't designed for that kind of performance.

At this point, many engineers in the audience would like to get up and knock this guy's block off. Here we are, the heroic engineers, doing the calcs (with checking), improvising the fit of a round carbon dioxide scrubber in a square box, solving one insurmountable problem after another. On top of all of that, who do we have to put up with? A lawyer! The company lawyer is whining about liability with the astronauts this close to perishing. If, in fact, the lunar module doesn't perform since it wasn't designed that way, the lawyer has covered the company's tracks. Of course, you end up with a spaceship full of dead astronauts.

Apollo 13 depicted what we engineers do, albeit in a caricatured and exaggerated way. Our mission is noble and heroic, although in reality, our work story unfolds much slower and less dramatically than in the movie. But as in *Apollo 13,* lives depend on our successful work. Society is built on the infrastructure and systems that we design and construct. Society expects flawless performance from these systems, and we manage to deliver. Our bridges rarely fall down. The water flows from the tap when you turn the spigot. These things don't happen by themselves. Because these engineering successes happen regularly and with little fanfare, people take them for granted. When you walk in a building, you don't expect it to collapse. Civil engineering projects are expected to perform perfectly. The expectations are met so often that it's a great shock when there is a failure. Then we get plenty of publicity.

I've often wished that there were more popular presentations of what engineers do, like what was shown in *Apollo 13*. Our work is exciting, gratifying, and extremely important, and we're good at it. It's time for more engineering TV shows and movies. We've seen enough of *L.A. Law*. It's time for *Encino Engineer*.

Aberaeron

Driving north on highway A487 along the Welsh shore, you arrive at the village of Aberaeron. In the Welsh language, "aber" means "estuary," and this helps to explain why a lot of the village names along the coast begin with "aber." The drive into town is very picturesque and unexpected. You approach the town from a high bluff along the Irish Sea and then, with little warning, suddenly descend to the village. The town includes several groups of brightly painted, Victorian row houses gathered around a small, protected harbor. On that day we had intended to keep going another 15 miles or so to another, larger town (Aber-something), which was recommended by the tourist book. But when we arrived at Aberaeron, we found it instantly appealing and we decided to stop there for lunch.

It was a beautiful, warm, cloudless day, which apparently happens in Wales once every three years or so. We found a nice restaurant near the harbor and had lunch on a patio next to the water. On the far bank, the houses were all painted different bright colors, like something you'd see in Scandinavia. A native explained to us that this was a recent innovation, with the houses repainted only a few years ago. Before that, the colors were more drab and uniform. Overall, the harbor scene was tranquil and slow, befitting a vacation view. The River Aeron flowed into the harbor. The river wasn't much of a river, really a creek. An attractive timber bowstring arch bridge for pedestrians crossed the river mouth. It was a nice little bridge, but in a more modern style not quite in keeping with the quaint, historical theme of the rest of the village. We later learned that Aberaeron was a new town by Welsh standards, only around since the 18th century. In the United States, of course, a town from the

18th century would be considered ancient and good for some historic plaques and mention on the register.

After lunch, we walked down a promenade next to the harbor. Two energetic teenage boys in wet suits were busy jumping off the retaining wall into the water and climbing back up with bravado. I looked a bit farther to find the source of the bravado, and sure enough, two teenage girls were egging them on. The harbor water was fed by the Irish Sea and it was cold, but the boys wanted to show off and, fortunately, they did have wet suits on. The promenade was protected from the harbor by a separate seawall set back from the retaining wall. The structure had openings that you had to climb over to cross to the space next to the water. This detail suggested flood protection for the adjacent shops—definitely the weather wasn't always as placid as the day of our visit.

We reached the end of the harbor promenade, which took a 90-degree bend to the north and became a seafront promenade. A young boy, maybe six or seven years old, approached my teenage son Dan and started peppering him with questions. Dan was chewing gum, so the boy said:

"What's that in your mouth?" His accent was a combination of boy/Welsh/British, with an upturn in tone at the end of each phrase.

Dan said: "Gum."

"What's gum?" asked the boy.

"You chew on it," Dan said.

"Why do you chew on it?" asked the boy.

"Because it's good to."

"Do you like chewing on it?" asked the boy.

"Yes," said Dan.

"What's it made out of?" asked the boy.

Dan thought about that one for a second. "It's made out of gum."

"Oh," said the boy.

The conversation went on for about 15 minutes or so. It was an interesting face-off between the boy's fearless precocity and my son's teenage reticence. Two adults watched the conversation with interest from a short distance: the teenager's father and the boy's mother, who wore a smile suggesting a combination of bemused pride and slight concern that her son was being a pest. Probably there is some interesting word for "pest" in Welsh (such as "poendod"). It would be a word that could be pronounced once you got used to it, but that would be virtually impossible to spell. Driving around Wales, there were signs with many words of this type above the more recognizable English translations.

Sunlight sparkled on the sea. To the north, beyond the promenade, a distant, misty bluff jutted out into the water, forming Cardigan Bay. It looked welcoming but also a little cold and forbidding. On the other side of the sea, you could just make out the green outline of Irish headlands. I was told that on an extraordinarily clear day, maybe once every eight years, you could see Ireland in more detail.

The slow and peaceful town showed few signs of any past trauma, but on that calm, sunny day there were signs if you looked closely. The seawall along the Irish Sea had human-made, rocky outcrops jutting into the water and obstructing the shoreline. These were probably placed in defense of seaside erosion during a storm, but my imagination wandered and I thought of a different type of defense. Perhaps the boulders had been placed there as obstacles for invading boats during World War II. In that case, even in Aberaeron, about as far from the rest of the world as possible, the rest of the world wasn't that far away. For a moment, I thought of the novel, *On the Beach*. In that genteel, pleasant doomsday story, the Australians peacefully go about their business isolated from disaster until disaster relentlessly creeps up the coast. Today in the United States, one form of threat had been addressed by plopping concrete Jersey barriers on sidewalks in front of buildings. After the initial flurry died down, the haphazardly plopped Jersey barriers were replaced by more thoughtfully detailed architectural bollards, retrofitted onto the sidewalk plazas. Unlike the Jersey barriers, the bollards seem to blend into the sidewalks, but not completely. They don't appear to serve any function, except in the context of addressing the threat. I wondered if future pedestrians will look at the bollards the same way I looked at the rock outcrops in the water. Would they see the bollards as physical vestiges and reminders of a past threat? Or would the bollards still be needed?

When the young boy was finally satisfied, the conversation ended, and Dan and my daughter Rachel climbed down some steps along the seawall to a small beach. They skipped stones. Dan found some flat stones and made some great throws, maybe eight or nine skips. The flat rocks bounced across the shimmering surface. Light bathed and blended into everything. You lost sight of the skipping stones and could imagine that they were skipping all the way to Ireland. When it was time to move on, we loaded into the car and traveled once again on what seemed like the wrong side of the road. I asked Dan:

"What's that in your mouth?"

"Gum," he said.

"What's gum?" I asked.

I received a good glare for his response. That concluded our conversation for many miles. We drove on, mesmerized by the greens and blues of the Welsh countryside. The village of Aberaeron faded into the shoreline mist.

Publishing Credits

The following essays first appeared in *Tufts Journal* and are reproduced with permission from Tufts University: "Brian's Bridges," "Who Likes the Chocolate?" "The Baby Sitter-in-Law," and "My New Cell Phone."

The following essays first appeared in *PB Network,* the internal technical journal of Parsons Brinckerhoff, and are reproduced with permission: "The Twister," "Don't Throw This Away," "Don't Throw This Away, II" (originally titled, "Don't Throw This Away Either"), "Don't Throw This Away, III" (originally "Don't Throw This Away, Part IV"), "Don't Throw This Away, IV" (originally "Don't Throw This Away, Part V"), "New Car," "Acronyms and the Explosion of Useless Data (AEUD)," "Fish," "Fred Retires," "A Comparison of Dilbert and Wally," "The Forest and the Trees," and "Encino Engineer."

The following essays first appeared in *Civil Engineering Practice* and are reproduced with permission from the Boston Society of Civil Engineers Section: "Engineering Fashions," "The Discovery of Pluto," "What's New on the Xway" (originally titled "Enjoying the View"), "The Last Game at Foxboro," "Moss on the Median," "What Happened to Nantucket?" and "Mass MoCA and the Hoosac Tunnel."

The following essays first appeared in *Journal of Leadership and Management in Engineering* and are reproduced with permission from ASCE: "Life Insurance," "The Maze," "The Ronald Reagan Room," "Bringing Out the Inner Civil Engineer," "Learning the Expanding Body of Knowledge," "The Road Not Built," "Built It and They Will Come," "Raising the Bar," "Hamsters Gone Wild," "Opryland," "The Zucchini Story," "First Class," "The Way Things Are," and "The Trail Ridge Road."

The following essay first appeared in the *Journal of Professional Issues in Engineering Education and Practice* and is reproduced with permission from ASCE: "An Ideal Geotechnical World."

About the Author

Brian Brenner, P.E., got his start in civil engineering at age three, building suspension bridges out of blocks. With his dad driving, he was the first toddler to cross the Verrazano-Narrows Bridge on its opening day. Later in life, he was glad to have two great children so he could play with blocks again and nobody would think it was strange, including his beautiful and often bemused wife, Lauren.

Professionally, he teaches classes at Tufts University in concrete design, bridge analysis and design, bridge history, aesthetics, and introduction to engineering. His research includes long-term bridge design, structural parameter estimation, and topics in engineering education. Prior to his appointment at Tufts, he was a practicing structural engineer with Parsons Brinckerhoff in Boston. He has published more than 70 papers and articles on structural analysis and design, design for construction mitigation, engineering education, computer-aided design, and other topics. He is editor emeritus of *Journal of Professional Issues in Engineering Education and Practice,* the education journal of the American Society of Civil Engineers, and editor of the ASCE journal, *Leadership and Management in Engineering*. He is chair emeritus of the publications committee of *Civil Engineering Practice,* the journal of the Boston Society of Civil Engineers section (BSCES), and he is active in several BSCES and ASCE committees. He received the BSCES President's Award in 2000, the Clemens Herschel Award in 2001, and the Richard R. Torrens Award from ASCE in 2005.